Anna Unkovskaya

Assessment of leak frequency of cross-country oil pipelines

Anna Unkovskaya

Assessment of leak frequency of cross-country oil pipelines

LAP LAMBERT Academic Publishing

Impressum / Imprint

Bibliografische Information der Deutschen Nationalbibliothek: Die Deutsche Nationalbibliothek verzeichnet diese Publikation in der Deutschen Nationalbibliografie; detaillierte bibliografische Daten sind im Internet über http://dnb.d-nb.de abrufbar.

Alle in diesem Buch genannten Marken und Produktnamen unterliegen warenzeichen-, marken- oder patentrechtlichem Schutz bzw. sind Warenzeichen oder eingetragene Warenzeichen der jeweiligen Inhaber. Die Wiedergabe von Marken, Produktnamen, Gebrauchsnamen, Handelsnamen, Warenbezeichnungen u.s.w. in diesem Werk berechtigt auch ohne besondere Kennzeichnung nicht zu der Annahme, dass solche Namen im Sinne der Warenzeichen- und Markenschutzgesetzgebung als frei zu betrachten wären und daher von jedermann benutzt werden dürften.

Bibliographic information published by the Deutsche Nationalbibliothek: The Deutsche Nationalbibliothek lists this publication in the Deutsche Nationalbibliografie; detailed bibliographic data are available in the Internet at http://dnb.d-nb.de.

Any brand names and product names mentioned in this book are subject to trademark, brand or patent protection and are trademarks or registered trademarks of their respective holders. The use of brand names, product names, common names, trade names, product descriptions etc. even without a particular marking in this work is in no way to be construed to mean that such names may be regarded as unrestricted in respect of trademark and brand protection legislation and could thus be used by anyone.

Coverbild / Cover image: www.ingimage.com

Verlag / Publisher:
LAP LAMBERT Academic Publishing
ist ein Imprint der / is a trademark of
OmniScriptum GmbH & Co. KG
Heinrich-Böcking-Str. 6-8, 66121 Saarbrücken, Deutschland / Germany
Email: info@lap-publishing.com

Herstellung: siehe letzte Seite /
Printed at: see last page
ISBN: 978-3-659-44965-9

Copyright © 2015 OmniScriptum GmbH & Co. KG
Alle Rechte vorbehalten. / All rights reserved. Saarbrücken 2015

TABLE OF CONTENTS

1. INTRODUCTION 1
2. PROCEDURE OF CROSS-COUNTRY OIL PIPELINES ACCIDENTAL LEAK FREQUENCY CALCULATION 4
REFERENCES..12
APPENDIX 1. ANALYSIS OF AVAILABLE STATISTIC DATA ON ACCIDENTS AT CROSS-COUNTRY OIL PIPELINES IN RUSSIA AND ABROAD 14
APPENDIX 2. U.S. DOT STATISTIC DATA ANALYSIS – PHMSA.......................................22
APPENDIX 3. PROCEDURE DEVELOPMENT FOR ACCIDENTAL LEAK FREQUENCY CALCULATION 39
APPENDIX 4. AN EXAMPLE OF ACCIDENTAL LEAK FREQUENCY OF OP CALCULATION 67
APPENDIX 5 (FOR REFERENCE). BRIEF DESCRIPTION OF ANALYZED ACCIDENTS CONSEQUENCES: FIRES AND EXPLOSIONS 75

1. INTRODUCTION

Assessment of the expected initiating events frequency onsite may be based on a historical analysis of accident rates at similar objects. The statistical analysis should identify ("feel") essential regularities in the accidents frequency and range for their subsequent application, when calculating the accidents occurrence frequency onsite. For the most reliable assessment of initiating events frequency, it is important that statistical data on already occurred accidents enable identification of essential factors influence (natural, man-made, technological) typical for a specific object on the frequency of accidents occurrence, i.e. are representative. Cross-country oil pipelines (OP) represent the class of objects convenient for such analysis, as the large amount of statistical data on OP accidents collected and systematized in special databases has been accumulated worldwide.

The author performed comparative analysis of accessible statistical databases on OP accidents presented in Appendix 1. The analysis included well-known statistical databases of various organizations in Europe and North America with publicly available accumulated information. The analysis highlighted the most appropriate statistical database for calculation of OP accidental leak frequency with the best possible accounting of implemented safety measures, as well as causes and factors contributed to accidents occurrence.

As a result of analysis, the database was chosen managed by the Pipelines and Hazardous Materials Safety Administration (PHMSA) of the US Department of Transportation (DOT). Accumulated statistical data of this database have been subjected to detailed study (the total of 1,327 accidents that occurred in the period 1984–2013).

Basic principles of analysis, processing and adaptation of selected statistical data are given in detail in Appendix 2. The data are presented on accidents distribution depending on causes of their occurrence in different analyzed periods (decades) (1984–2013 and 2004–2013). OP exposure was identified: total exposure, depending on the pipeline diameter, presence/absence of cathodic protection and/or corrosion-resistant coating (for the period 1984–2013), as well as depending on the year of OP construction (for the period 2004–2013).

Appendix 3 contains information on the calculation procedure development for expected accidental leak frequency of OP on each section. It presents the rationale for choosing calculation formulas, ratios and recommended values of correction factors, which correct the value of primary frequency in calculations. The possibility exists of adjustment and/or assignment of additional correction factors depending on the availability of expert appraisal and/or experimental (practical) information on the impact of various factors on the safety level of specific OP.

It is presented the accidental leak frequency distribution of OP by six classes of causes: external interference, corrosion, construction/material defect, natural forces, incorrect operation, other and unknown, – as well as depending on the nature of pipeline damage.

Partially the first version of the specified procedure was published in 2014 in the "Bezopasnost Zhiznedeyatelnosti" (Life Safety) magazine [1, 2], where all analyzed accidents were included for the period 1984–2011. Primary accidental leak frequencies of OP, calculation formulas of correction factors, leak frequencies and full ruptures (100 % cross section) in case of OP accidental leak, were slightly changed in this version.

The proposed procedure is similar to the calculation procedure of accidental leak frequency of cross-country gas pipelines (GP) [3], developed on the basis of the statistical database of the European Gas Pipeline Incident Data Group (EGIG) [4] and included in the Project Specific Technical Specification Quantitative Risk Assessment for hazardous industrial facilities of the "Sakhalin-II" Project [5].

The example of calculation of a section of one modern OP is presented in Appendix 4. The calculation results showed "sensitivity" of the proposed procedure to provided safety measures and to the variety of conditions of OP routing.

Appendix 5 serves as reference, since the information presented in it concerns the analysis of analyzed accidents (fires and explosions) consequences, but the author considers appropriate to publish these data.

2. PROCEDURE OF CROSS-COUNTRY OIL PIPELINES ACCIDENTAL LEAK FREQUENCY CALCULATION

To perform calculation, the OP route is divided into sections with calculation of corresponding accidental leak frequency for each section. Division into sections can be performed depending on design and engineering solutions affecting the OP safety (e.g. diameter, wall thickness or pipeline class, pipeline laying method, etc.), and/or depending on climatic conditions of pipeline routing (e.g. presence of tectonic faults, water or common roads crossings, etc.). Furthermore, the division may correspond to sections taken in the design or engineering construction documentation.

The estimated accident frequency F in any OP section is calculated by the formula:

$$F = \sum_{i=1}^{S} f_i \qquad (1)$$

where f_i – accidental leak frequency for the *i-th* cause class (1/1,000 km·yr);

S – number of classes of leak causes (where S = 6, Fig. 1).

For each OP section, the final accidental leak frequency based on OP damage size is calculated by the formula:

$$F_k(m) = \sum_{i=1}^{S} f_{ik}(m) \qquad (2)$$

where $F_k(m)$ – accidental leak frequency for the *k-th* damage size of the *m-th* section of OP (1/1,000 km·yr);

$f_{ik}(m)$ – accidental leak frequency for the *k-th* damage size of the *i-th* class of accidental leak causes of the *m-th* section of OP (1/1,000 km·yr);

$i = \{1...6\}$ – classes of accidental leak causes of OP;

$k = \{1...2\}$ or $\{1...5\}$ – types (sizes) of damages: leak / full rupture (100 % cross section) $\{1...2\}$ or diameters (mm) of equivalent emergency openings: 12.5; 25; 50; 100; > 150 – rupture $\{1 ... 5\}$.

When $k = \{1, 2 ...\}$, the expected leak frequency of the analyzed OP section is calculated by the formula:

$$F_L(m) = 0{,}8714 \cdot \sum_{i=1}^{S} f_i \qquad (3)$$

The expected frequency of full ruptures (100 % cross section) of the analyzed OP section is calculated by the formula:

$$F_R(m) = 0{,}1286 \cdot \sum_{i=1}^{S} f_i \qquad (4)$$

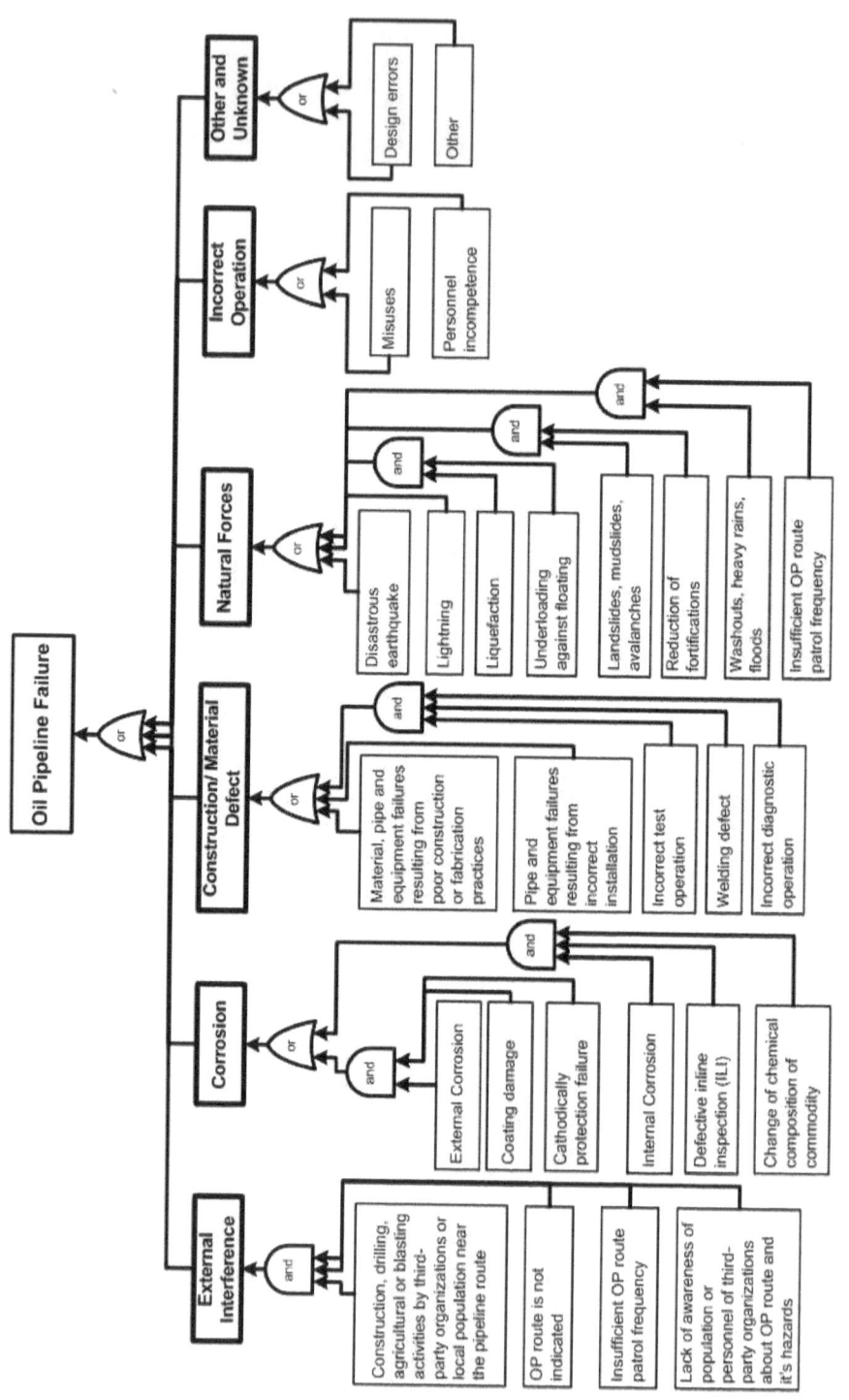

Figure 1 – Fault tree at OP accidental leak

Table 1 shows the formulas for OP accidental leak frequencies calculation depending on the class of accident causes. Conventions used in Table 1 are as follows:

f_C	OP accidental leak frequency by the cause class "Corrosion";
f_{PC}	Primary frequency of OP accidental leak by the cause class "Corrosion";
f_{EI}	OP accidental leak frequency by the cause class "External interference";
f_{PEI}	Primary frequency of OP accidental leak by the cause class "External interference";
f_{EID}	OP accidental leak frequency by the cause class "External interference" calculated in accordance with OP diameter;
f_{CD}	OP accidental leak frequency by the cause class "Construction/material defect";
f_{PCD}	Primary frequency of OP accidental leak by the cause class "Construction/material defect";
f_{NF}	OP accidental leak frequency by the cause class "Natural forces";
f_{PNF}	Primary frequency of OP accidental leak by the cause class "Natural forces";
f_{BNF}	Background accidental leak frequency of OP by the cause class "Natural forces" caused by geological hazards of an OP route that are not accounted in the basic formula;
f_{SA}	Accidental leak frequency of OP caused by seismic action;
$T(2PGA)$	Frequency of earthquakes recurrence, characterized by double excess of the level of peak ground acceleration (PGA) typical for analyzed region / area of a pipeline route;
f_{TF}	Accidental leak frequency of OP caused by its destruction on the active tectonic fault that is defined by taking into account the earthquake return period in the area of OP route, resulting in the pipeline damage in the area of tectonic fault;
f_L	Accidental leak frequency of OP caused by landslides and mudslides;
f_{LF}	Accidental leak frequency of OP caused by seismic soil liquefaction;
f_{WC}	Accidental leak frequency of OP caused by soil movement at OP water crossings;
f_S	Accidental leak frequency of OP at swampy areas crossing;
f_{IO}	OP accidental leak frequency by the cause class "Incorrect operation";
f_{PIO}	Primary frequency of OP accidental leak by the cause class "Incorrect operation";
f_{IOD}	OP accidental leak frequency by the cause class " Incorrect operation" calculated in accordance with OP diameter;
f_{OU}	OP accidental leak frequency by the cause class "Other and Unknown";
f_{POU}	Primary frequency of OP accidental leak by the cause class "Other and Unknown";
k_{WT}	Correction factor of accidental leak frequency of OP, taking into account

	the effect of OP wall thickness;
k_{CP}	Correction factor of accidental leak frequency of OP, taking into account the use of corrosion protection measures;
k_D	Correction factor of accidental leak frequency of OP, taking into account the effect of OP diameter;
k_{HDD}	Correction factor of accidental leak frequency of OP, taking into account the OP laying, by using the method of horizontal directional drilling;
k_C	Correction factor of accidental leak frequency of OP, taking into account the effect of OP railways, roads and utilities crossings;
l_j	Total length of OP sections in the analyzed area with safety classes, with the wall thickness less than 10 mm;
l_q	Total length of OP sections in the analyzed area with safety classes, with the wall thickness more than 10 mm;
l_C	Length of OP railways, roads and utilities crossings in the analyzed area;
l_{SEC}	Total length of OP analyzed section;
l_{CR}	Length of OP categorized roads and railways in the analyzed area, it is recommended to take 25 m;
l_R	Length of OP other roads crossing in the analyzed area, it is recommended to take 20 m;
l_U	Length of OP utilities crossing in the analyzed area, it is recommended to take 5 m;
l_{TF}	Length of OP active tectonic fault crossing in the analyzed area, which estimated length shall be equal to the size of the fault uncertainty zone;
T	Earthquakes recurrence period in the area of OP routing, causing the pipeline damage in the tectonic fault area;
$l_{L(LF)}$	Length of OP crossing areas with possible formation of landslides, mudslides, seismic soil liquefaction;
l_{WC}	Length of OP water crossings in the analyzed area;
l_S	Length of OP swampy areas crossing in the analyzed area;
(1)	Primary (average) accidental OP leak frequency calculated for 30-year period (1984–2013);
(2)	Primary (average) accidental OP leak frequency calculated for 10-year period (2004–2013);
(3)	Primary (average) accidental OP leak frequency calculated for 10-year period (2004–2013, including small spills).

Table 1 – Calculation formulas of accidental leak frequency of OP per accidents cause class

Cause class	Basic calculation formula	Primary frequency, (1/1000 km.yr)	Subsidiary calculation formulas and values of correction factors
Corrosion	$f_C = f_{PC} \cdot k_{WT} \cdot k_{CP}$	(1) $2.7 \cdot 10^{-1}$ (2) $1.88 \cdot 10^{-1}$ (3) $4.12 \cdot 10^{-1}$	Given the length of pipeline sections of various safety classes in the analyzed area: $$k_{WT} = \frac{1 \cdot \sum_{j=1}^{N} l_j + 0.018 \cdot \sum_{q=1}^{M} l_q}{\sum_{l=1}^{N} l_j + \sum_{q=1}^{M} l_q}$$ k_{WT}=0.018, if the OP wall thickness is more than 10 mm, k_{WT}=1, if the OP wall thickness is less than or equal to 10 mm.	k_{CP}=0.6, if cathodically protected with protective coating, k_{CP}=7.4, if cathodically protected without protective coating, k_{CP}=10.6, if cathodically unprotected without protective coating, k_{CP}=4, if cathodically unprotected with protective coating.
External interference	$f_{EI} = f_{PEI} \cdot k_D \cdot k_{HDD} \cdot k_C$	(1) $1.76 \cdot 10^{-1}$ (2) $9.0 \cdot 10^{-2}$ (3) $1.17 \cdot 10^{-1}$	$k_{WT} = f_{EID}/f_{PEI}$, where $f_{EID} = 0.6951 \cdot \exp(-0.1 \cdot D)$ Given the length of pipeline crossings: $k_C = 1 \cdot \left(1 - \frac{l_C}{l_{SEC}}\right) + 2 \cdot \frac{l_C}{l_{SEC}}$, $l_C = l_{CR} + l_R + l_U$	k_{HDD}=0, if the area is performed using HDD, k_{HDD}=1 in other areas. k_C=2, in the area with roads, railways, utilities crossing, k_C=1, in other areas.
	For a pipeline of the same diameter with different safety classes (different wall thickness): $f_{EI} = f_{PEI} \cdot k_{WT} \cdot k_{HDD} \cdot k_C$		Given the length of pipeline sections of various safety classes in the analyzed area: $$k_{WT} = \frac{1 \cdot \sum_{j=1}^{N} l_j + 0.022 \cdot \sum_{q=1}^{M} l_q}{\sum_{l=1}^{N} l_j + \sum_{q=1}^{M} l_q}$$ Given the length of pipeline crossings:	k_{HDD}=0, if the area is performed using HDD, k_{HDD}=1 in other areas. k_C=2, in the area with roads, railways, utilities crossing, k_C=1, in other areas. k_{WT}=0.022, if the OP wall thickness is more than 10 mm, k_{WT}=1, if the OP wall thickness is less than or equal to 10 mm.

Cause class	Basic calculation formula	Primary frequency, (1/1000 km·yr)	Subsidiary calculation formulas and values of correction factors	
Construction /material defect	$f_{CD} = f_{PCD}$	(1) $9{,}1 \cdot 10^{-2}$ (2) $7{,}3 \cdot 10^{-2}$ (3) $1{,}94 \cdot 10^{-1}$	$k_C = 1 \cdot \left(1 - \dfrac{l_C}{l_{SEC}}\right) + 2 \cdot \dfrac{l_C}{l_{SEC}};$ $l_C = l_{CR} + l_R + l_U$	–
Natural forces	$f_{NF} = f_{BNF} + f_{SA} + f_{TF} + f_L +$ $+ f_{LF} + f_{WC} + f_S$	(1) $2{,}5 \cdot 10^{-2}$ (2) $1{,}4 \cdot 10^{-2}$ (3) $3{,}1 \cdot 10^{-2}$	$f_{BNF} = 0{,}352 \cdot f_{PNF} \cdot k_{WT}$, где given the length of pipeline sections of various safety classes in the analyzed area: $k_{WT} = \dfrac{1 \cdot \sum\limits_{j=1}^{N} l_j + 0{,}22 \cdot \sum\limits_{q=1}^{M} l_q}{\sum\limits_{l=1}^{N} l_j + \sum\limits_{q=1}^{M} l_q}$	$k_{WT}=0{,}22$, if the OP wall thickness is more than 10 mm, $k_{WT}=1$, if the OP wall thickness is less than or equal to 10 mm.
			$f_{SA} = \left(1/\{T(2PGA)\cdot 10\}\right) \cdot k_{WT}$, где given the length of pipeline sections of various classes of safety in the analyzed area: $k_{WT} = \dfrac{1 \cdot \sum\limits_{j=1}^{N} l_j + 0{,}022 \cdot \sum\limits_{q=1}^{M} l_q}{\sum\limits_{l=1}^{N} l_j + \sum\limits_{q=1}^{M} l_q}$	$k_{WT}=0{,}022$, if the OP wall thickness is more than 10 mm, $k_{WT}=1$, if the OP wall thickness is less than or equal to 10 mm.
			$f_{TF} = \dfrac{1}{T \cdot l_{TF}}$; $f_L = f_{LF} = 0{,}1 \cdot f_{SA} \cdot \dfrac{l_{LCLF}}{l_{SEC}}$	$f_S = 2 \cdot f_{BNF} \cdot \dfrac{l_S}{l_{SEC}}$
			If the area is performed using HDD method, then $f_{WC} = f_{BNF} \cdot \dfrac{l_{WC}}{l_{SEC}}$	In other areas: $f_{WC} = 5 \cdot f_{BNF} \cdot \dfrac{l_{WC}}{l_{SEC}}$

Cause class	Basic calculation formula	Primary frequency, (1/1000 km.yr)	Subsidiary calculation formulas and values of correction factors
Incorrect operation	$f_{IO} = f_{PIO} \cdot k_D$	(1) $1,1 \cdot 10^{-2}$ (2) $1,3 \cdot 10^{-2}$ (3) $2,1 \cdot 10^{-2}$	$k_D = f_{IOD} / f_{PIO}$, where $f_{IOD} = -0,019 \cdot \ln(D) + 0,0681$
Other and Unknown	$f_{OU} = f_{POU}$	(1) $5,2 \cdot 10^{-2}$ (2) $1,3 \cdot 10^{-2}$ (3) $3,6 \cdot 10^{-2}$	-

REFERENCES

1. Unkovskaya A.V. Assessment of Leak Frequency of Cross-Country Oil Pipelines (Part 1) // Bezopasnost' zhiznedejatel'nosti (Life Safety). – 2014. – №11. – P.60-72.
2. Unkovskaya A.V. Assessment of Leak Frequency of Cross-Country Oil Pipelines (Part 2) // Bezopasnost' zhiznedejatel'nosti (Life Safety). – 2014. – №12. – P.45-60.
3. Shavkin S.V., Chernoplekov A.N., Gosteva A.V., Monakhov R.E., Ljapin A.A. Accident leak frequency calculation algorithm for quantitative risk assessment of modern cross-country gas pipelines // Bezopasnost' zhiznedejatel'nosti (Life Safety), prilozhenie k zhurnalu (Appendix). – 2009. – №3. – P.1-24.
4. EGIG; European Gas pipeline Incident data Group; 6th EGIG report 1970-2004; EGIG document 05.R.0002, issued December 2005 (www.EGIG.nl).
5. Addendum to Project Specific Technical Specification (PSTS) Quantitative Risk Assessment for hazardous industrial facilities of the "Sakhalin-II" Project "Onshore Gas Pipelines" (Doc. № 5600-C-90-04-S-1001-00). Moscow, January 2007.
6. CONCAWE, Performance of European cross-country oil pipelines, Statistical summary of reported spillages in 2011 and since 1971, Report No. 3/13, CONCAWE, Brussels, April 2013.
7. NEB, National Energy Board Canada, Focus on Safety and Environment. A Comparative Analysis of Pipeline Performance 2000 – 2009, December 2011 (http://www.neb-one.gc.ca/).
8. UKOPA, Pipeline Product Loss Incidents (1962-2004), 4th Report of the UKOPA Fault Database Management Group, Report Number R 8099, April 2005.
9. Annual reports and accidents/incident data for hazardous liquids of Pipeline and Hazardous Materials Safety Administration (PHMSA) Department of Transportation (DOT) U.S. (http://phmsa.dot.gov/pipeline).

10. A Guideline "Using or Creating Incident Databases for Natural Gas Transmission Pipelines". Report of Study Group 3.4. 23rd World Gas Conference June 1–5, 2006 Amsterdam, the Netherlands.
11. Forms. Accident/Incident/Annual Reporting Forms [Digital source] // PHMSA. URL: http://www.phmsa.dot.gov/pipeline/library/forms. (Date of access: 02/23/2015).
12. Distribution, Transmission and Liquid Accident and Incident Data [Digital source] // PHMSA. URL: http://phmsa.dot.gov/portal/site/PHMSA/menuitem.ebdc7a8a7e39f2e55cf20310 50248a0c/?vgnextoid=fdd2dfa122a1d110VgnVCM1000009ed07898RCRD&vg nextchannel=3430fb649a2dc110VgnVCM1000009ed07898RCRD&vgnextfmt= print. (Date of access: 02/23/2015).
13. Distribution, Transmission and Liquid Annual Data [Digital source] // PHMSA. URL: http://phmsa.dot.gov/portal/site/PHMSA/menuitem.ebdc7a8a7e39f2e55cf20310 50248a0c/?vgnextoid=a872dfa122a1d110VgnVCM1000009ed07898RCRD&vg nextchannel=3430fb649a2dc110VgnVCM1000009ed07898RCRD&vgnextfmt= print. (Date of access: 02/23/2015).
14. ERM Group Inc., QRA Report Onshore Facilities. Sakhalin 1 Project, Phase I., Annex D, October 2002 – P.19.
15. Set of rules SP 14.13330.2011 Construction in seismically active regions. Revised edition of SNiP II-7-81* / Minregion of RF. – M., 2011.
16. Zhulina S.A., Lisanov M.V., Savina A.V. Methodical guide on assessment of risk accident level at oil trunk pipelines and main oil products pipelines // Bezopasnost' truda v promyshlennosti. – 2013. – P.50–55.

APPENDIX 1.
ANALYSIS OF AVAILABLE STATISTIC DATA ON ACCIDENTS AT CROSS-COUNTRY OIL PIPELINES IN RUSSIA AND ABROAD

Collection of data on accidents at OP in the Russian Federation is performed by Rostechnadzor (Federal Environmental, Industrial and Nuclear Supervision service of Russia). According to the Federal Law No. 116-FZ, "On Industrial Safety of Hazardous Industrial Assets" dated 07/21/1997, incidents and accidents are recorded. These data are not publicly distributed. Since 2004, Rostechnadzor publishes annual reports on its work, where only general data on the level of accidents at the main pipeline transport are present. But they cannot be used for accidental leak frequency calculation of a particular OP. Large Russian companies – operating OP, e.g. JSC AK Transneft, JSC Lukoil also keep their internal databases on accidents that are not published.

To date, sufficient statistical data on incidents/ accidents at OP is accumulated worldwide, various organizations have their own databases (Table A.1-1).

Table A.1-1 – Organizations managing databases on incidents / accidents at OP

Organizations	Country/Region	Official Website
International Association of Oil and Gas Producers (OGP)	Great Britain	http://www.ogp.org.uk/
Pipeline and Hazardous Materials Safety Administration (PHMSA), Department of Transportation (DOT)	USA	http://www.dot.gov http://phmsa.dot.gov/pipeline
United States Department of Labor – Bureau of Labor Statistics (BLS)	USA	http://www.bls.gov/
National Energy Board of Canada (NEB)	Canada	http://www.neb.gc.ca/
Human Resources and Skills Development Canada (HRSDC)	Canada	http://www.esdc.gc.ca/
Canadian Association of Petroleum Producers (CAPP)	Canada	http://www.capp.ca/
BC Oil and Gas Commission (OGC)	Canada	http://www.bcogc.ca/
Alberta Energy and Utilities Board (AER)	Canada	http://www.aer.ca/
Pipe Line Contractor Association of Canada (PLCAC)	Canada	http://www.pipeline.ca/
European Oil Companies Association for Environment, Health and Safety	Europe	https://www.concawe.eu/

Organizations	Country/Region	Official Website
(CONCAWE)		
United Kingdom Onshore Pipeline Operators' Association (UKOPA)[1]	Great Britain	http://www.ukopa.co.uk/
Rostechnadzor (Federal Environmental Industrial and Nuclear Supervision service of Russia)	Russian Federation	http://www.gosnadzor.ru/
The Australian Pipeline Industry Association Ltd (APIA)	Australia	http://www.apia.net.au/

For the purpose of this paper, the author carried out detailed comparative analysis of available in the public media databases on incidents / accidents at OP, performed on the basis of works [6–9], taking into account materials and reports from official websites of specified organizations (Table A.1-1), which main results are shown below. Forms of tables were taken according to [3, 7, 10].

Criteria used for incidents / accidents identification in different databases are presented in Tables A.1-2, A.1-3 and A.1-4.

Basic characteristics of OP that should be met by OP to be included in the database are shown in Table A.1-5.

Table A.1-2 – Databases scope

Database	Ruptures Causes	Injury Frequency	Liquid releases, Leaks and Spills	Gas Releases
CONCAWE	-	-	X	-
PHMSA	X	-	X	X
BLS	-	X	-	-
NEB	X	X	X	X
AER	X	-	X	X
CAPP	-	X	-	-
PLCAC	-	X	-	-
OGP	-	X	-	-
HRSDC	-	X	-	-
Rostechnadzor	X	X	X	X

---

[1] UKOPA database basically contains information related to GP. According to the report [8] of this organization at the end of 2004, the total length of the reporting cross-country pipelines was 21,727 km, of them 20,001 km – GP, and 212.6 km – OP. Furthermore, the data in UKOPA reports concerning the number of accidents, their distribution depending on causes of occurrence and other information is presented for all reporting pipelines, regardless of transported product. Therefore, this database was excluded from further analysis.

Table A.1-3 – Incidents / accidents identification in databases by consequences

Database	Consequences				
	Unintentional Commodity Release	Fatality and Injury	Fire or Explosion	Environmental Impact	Other
CONCAWE	X	X	X	X	-
PHMSA	X	X	X	-	Emergency shutdown, significant Incident
BLS	-	X	-	-	-
NEB	X	X	X	X	Pipeline operated beyond design specifications
AER	X				Incidents without commodity release
CAPP	-	X	-	-	-
PLCAC	-	X	-	-	-
OGP	-	X	-	-	-
HRSDC	-	X	-	-	-
APIA	X	-	-	-	Damage or Defects requiring repair or MAOP derating, Near Misses
Rostechnadzor	X	X	X	-	-

Table A.1-4 – Databases reporting criteria for incidents/accidents

Database	Reporting Requirements
CONCAWE	The minimum reportable spillage size has been set at 1 m³ (unless exceptional safety or environment consequences are reported for a < 1 m³ spill), onshore cross-country pipelines (including short estuary or river crossings but excluding under-sea pipeline systems)
PHMSA	Prior to January 2002: any failure/leakage in a pipeline, if the minimum spill volume was 50 (8 m³) and more barrels of hazardous liquid, for high volatile substances – 5 or more barrels. Regardless of the spill volume, if a fire or explosion, death or personal injury requiring hospitalization took place, if the expected property damage exceeds 5,000 USD (up to 1993) / 50,000 USD (since 1994). Prior to 1993, the "expected property damage" meant only the property of operator / third parties or of both of them simultaneously. Since 1994, the "expected property damage" of the property of operator / third parties or of both of them simultaneously, was including cost of collection and clean-up, as well as cost of lost product. Since February 2002: any failure/leakage in a pipeline, if the minimum spill volume was 0.1 (5 gallons) and more barrels of hazardous liquid, except that no report is required for a release of less than 5 barrels (0.8 m³) resulting from a pipeline maintenance activity. Regardless of the spill volume, if a fire or explosion, death or personal injury requiring hospitalization took place, if the expected property damage including cost of clean-up and recovery, value of lost product, and damage to the property of the operator or others, or both exceeds 50,000 USD.
NEB	Any unintended or uncontained release of liquid hydrocarbons. Leaks any

Database	Reporting Requirements
	volume (less than 1.5 m³ or larger) from operation of pipeline systems and arise from other components, releases more than 1.5 m³ from pipeline ("body" of pipe).
AER	Any leak or pipeline failure ("leak" means the escape of substance from a pipeline, "break" means a rupture in any part of a pipeline).
Rostechnadzor	Accident – destruction of constructions and/or technical devices used at hazardous industrial assets, uncontrolled explosion and/or release of hazardous substances (Federal Law No. 116-FZ dated 07/21/1997). Incident – failure or damage of technical devices used at hazardous production facility, discrepancy in the engineering process set mode (116-FZ dated 07/21/1997 as amended in the Federal Law No. 22-FZ dated 03/04/2013).

Table A.1-5 – Basic characteristics of supervised piping systems

Database	Pipeline System Scope				
	Installations Included	Pipe Material	Commodity	Onshore/ Offshore	Part of Chain
CONCAWE	Yes	Steel	Oil and Oil Products	Onshore	Transmission
PHMSA	Yes	Steel, Plastic, other	Natural Gas and Hazardous Liquids	Onshore and Offshore	Gathering, Transmission, Distribution
NEB	Yes	Steel	Natural Gas, Oil and other gases and liquids	Generally Onshore	Transmission, partially Gathering
AER	Yes	Steel, Plastic, other	Natural Gas, Oil and other gases and liquids	Onshore	Gathering, Transmission
APIA	Yes	All	Natural Gas, Oil and other gases and liquids	Onshore	Gathering, Transmission
Rostechnadzor	Yes	Steel	Natural Gas, Oil and other	Onshore and Offshore	Gathering, Transmission, Distribution

Data on possible causes of an incident / accident are shown in Table A.1-6.

Table A.1-6 – The main causes of incidents/accidents

Database	Corrosion	Construction/ material defects	External interference	Natural forces	Other Causes
CONCAWE	Internal, External	Seam weld, material or construction defects, lamination	Excavation damage, terrorism, theft, vandalism outside force damage	Temperature variations, landslide, subsidence, flooding and others	Incorrect operation
PHMSA	Internal, External, Stress-corrosion cracking	Material and seam weld defects, equipment failure	Excavation damage, other outside force damage (for example, explosion/fire of nearby industrial	Natural force damage, earth movement, lightning, heavy rains/floods, temperature, high winds	Incorrect operation, other

Database	Corrosion	Construction/ material defects	External interference	Natural forces	Other Causes
			assets damage by motorized vehicle/equipment not engaged in excavation, theft of commodity and others)		
NEB	Internal, External, Stress-corrosion cracking	Material defects, construction damage or girth weld failure	External interference	Natural force damage	Incorrect operation, other
AER	Internal, External, Stress-corrosion cracking	Material defects, connections failure, pipes or pipeline fitting damage, girth weld failure	External interference (земройные работы, вандализм)	Earth movement, lightning, flooding	Overpressure, installation errors, incorrect operation, other
APIA	Internal, External, Stress-corrosion cracking	Material defects, construction or design errors	External forces	Earth movement, lightning, soil erosion	Blasting, electrical surge or inducted voltage, other
Rostechnadzor	Internal, External, Stress-corrosion cracking	Weld defects	Third party interference	Earth movement, permafrost	Other

The OP failure mode at incident / accident is shown in Table A.1-7.

Table A.1-7 – Damage types of OP

Database	Pinhole/ Crack	Hole	Leak	Puncture	Full Rupture	Notice
CONCAWE	X	X	X	-	X	Length and width in mm are reported. Another types are reported: no hole (connection failures and etc.), split, fissure
PHMSA	X	-	X	X	X	Another types are reported: connection failure, tear/crack
NEB	-	-	X	-	X	
AER	-	-	X	-	X	
APIA	-	-	X	X	X	Gouge, deformation, coating damage, near miss
Rostechnadzor	-	-	X	X	X	

The availability of data collected for analysis (the degree of "openness") of databases is shown in Table A.1-8.

Table A.1-8 – Collected data availability

Database	Availability characteristic
CONCAWE	Publish periodical reports which present the results of statistical analyses of the data contained within their database. As well as a brief raw data is available.
PHMSA	Raw data is available online from the Pipeline and Hazardous Materials Safety Administration (PHMSA) website. A brief report for all the incidents is supplied yearly, categorizing the incidents by cause. Totals are also included for property damage, fatalities and injuries.
NEB, AER	Publish periodical reports which present the results of statistical analyses of the data contained within their database. They do not publish the raw data collected form databases.
APIA	Database is in development and is still being populated with data. Hence, no formal report is available to date.
Rostechnadzor	Since 2004, it publishes annual reports on the state of safety at supervised assets, including very brief data on accidents of cross-country pipelines.

Data pertaining to the extent of the database such as the total number of incidents/accidents and the total exposure of the pipelines (measures in kilometer years) is important to establish the reliability of the statistical data. The oldest statistics are PHMSA (с 1968 г.) and CONCAWE (с 1971 г.) which can show trends over several decades. In whole these statistics show a reduction in the frequency of oil leaks over the last years. The CONCAWE reports also give five year moving average values to highlight short term trends by filtering out older incidents/accidents. Also CONCAWE provides data of spill area, contamination water bodies (surface water and groundwater), spillage volume (m^3) (gross and net loss).

PHMSA and CONCAWE databases are of particular interest for the purpose of this publication.

CONCAWE specialists pay more attention to the volumes of spilled and collected oil products, areas of contamination, as well as to OP damage sizes, but they are not focused on the specifics of applied safety measures and natural features of OP routing that reduces the suitability of CONCAWE reports [6] for the use in the comprehensive assessment of potential accidents frequencies.

In contrast, the database of the US Department of Transportation (PHMSA) [9] contains very detailed description of accidents that allows adequate evaluation of certain safety measures used in the OP construction and operation. PHMSA database includes data on the occurred accidents since 1968. It is updated monthly and is publicly available in the PHMSA website [9]. The total length of OP in the United States, included in the database, was more than 90 thous. km in 2013.

The PHMSA advantages include: large volume of statistical observations; causes of pipeline damages are well documented and are amenable to detailed analysis; data on pipelines of various diameters are presented separately (causes of damages are considered separately as well).

All causes of accidents, presented in PHMSA database, are well documented and may be divided into 6 main classes (Table A.1-9).

Table A.1-9 – Accident cause classes and their characteristics

Accidents Cause Class	Description of damage causes in incidents / accidents
1. Construction /material defect	Damages caused by defects or anomalies within the material of the pipe body or within the pipe seam or other weld due to faulty manufacturing procedures, defects resulting from poor construction, installation, or fabrication practices, and in-service stresses such as vibration, fatigue and environmental cracking. And damages caused by malfunction of control/ relief equipment including valves, regulators or other instrumentation; failures of various types of connectors, connections and appurtenances, failures of the body of equipment or other material (including those caused by construction-, installation-, or fabrication-related and original manufacturing-related defects or anomalies), and all equipment-related failures.
2. Incorrect Operation	Damages resulting from operating, maintenance, repair or other errors by facility personnel, including but not limited to improper valve selection or operation, inadvertent overpressurization, or improper selection or installation of equipment.
3. Corrosion	Damages both of a pipeline and of pipeline fittings caused by internal, external corrosion; additionally may be specified causes of corrosion (stray currents, galvanic, atmospheric, microbiological and others).
4. Natural forces	Damages resulting from earth movement, earthquakes, landslides, subsidence, lightning, heavy rains/ floods, washouts, flotation, mudslide, scouring, temperature, frost heave, frozen components, high winds or similar natural causes.
5. External interference	Damages resulting from excavation damage by operator's personnel or by the operator's contractor or by the people or contractors not associated with the operator (third party). And releases/ failures resulting from not-excavation-related outside forces, such as nearby industrial, man-made, or

Accidents Cause Class	Description of damage causes in incidents / accidents
	other fire or explosion, damage by vehicles or other equipment, mechanical damage and intentional damage including vandalism, terrorism and theft commodity or equipment (other outside force damage).
6. Other and Unknown	Accidents whose cause is currently unknown, or where investigation into the cause has been exhausted and the final judgment as to the cause remains unknown, or where a cause has been determined which does not fit into any of the main cause categories listed.

APPENDIX 2. U.S. DOT STATISTIC DATA ANALYSIS – PHMSA

As stated above, PHMSA [9] publishes raw data on all accidents occurred at facilities that provide production, transportation, storage and handling of hazardous substances. The data refer to gathering lines, cross-country and distribution pipelines, both offshore and onshore, including pump stations, intermediate storage bases, tank farms, valve sites, which are intended to transport dangerous substances of gases and liquids, including crude oil, highly volatile liquids, oil products and refined products (diesel fuel, gasoline, kerosene, etc.), carbon dioxide, hydrocarbon (i.e. other non-toxic and not flammable substances). Although the data collection started since 1968, detailed and systematic data were available only since 1984. Available statistic data required thorough preliminary selection (see section A.2.1) and then their detailed statistical analysis.

The database is formed on the basis of reports of operating companies, which operate hazardous production facilities: the report on occurred incidents / accidents (according to the form *PHMSA F 7000-1 (revision 7-2014)*) and annual reports on the state of operated facilities (according to the form *PHMSA F 7000-1.1 (revision 6-2014)*), which are available in the website [11]. Requirements to the data structure, transmitted to PHMSA by an operating company, periodically were changed (upwards), but most noticeable changes occurred in the 2000s.

Currently, accident reports (the form *PHMSA F 7000-1*) include (but not limited to) very detailed information (Table A.2-1).

Table A.2-1 – Accident information

Reported information:	
Operator	Name, address, name and telephone number of the reporter
Accident	Local time and date of the event, location of the event (latitude and longitude), city/town, state, onshore/offshore
Part of system involved in accident	tank or storage vessel; terminal /tank farm equipment and Piping; pump/meter station equipment and piping; onshore pipeline, including valve sites, platform/deepwater port, including platform-mounted equipment and piping; offshore pipeline, including riser and raiser bend, year of installation/construction
Item involved in accident	Pipe (pipe body or seam), diameter, wall thickness, manufacturer, coating type; valve; pump; meter station; scraper/pig trap; sump/separator; flange; relief line; tubing; tank/vessel, other and etc.
Losses (Estimated)	Public/Community Losses: cost of emergency response phase, cost of

	Reported information:
	environmental remediation, other costs. Operator losses: value of product lost, value of operator property damage, other costs
Commodity	Name of commodity spilled (crude oil; HVLs/other flammable or toxic fluid which is a gas at ambient conditions; CO_2 or other non-flammable, non-toxic fluid which is a gas at ambient conditions; gasoline, diesel, fuel oil or other petroleum product which is a liquid at ambient conditions), estimated amount of commodity involved (barrels or gallons), amounts of spilled and recovered
Cause	On/Offshore Pipelines: type of leak or rupture, type of block valve used for isolation of immediate section, length of segment isolated, distance between valves
Environment	Area of accident: under pavement, underground, inside/under building, in open ditch, above ground, under water, other, depth of cover
Consequences	Number of operator employees/ contractor employees working for operator/ general public fatalities and injuries, product ignited, explosion, number of evacuation, elapsed time until area was made safe
Environmental impact	Wildlife impact, soil contamination, water contamination, anticipated remediation, long term impact assessment performed
Corrosion failure	Results of visual examination, location of corrosion, type of corrosion, cause of corrosion, pipe coating, cathodic protection (yes or no, year protection started), was the commodity treated with corrosion inhibitors or biocides
Natural forces	Earth movement (earthquake, subsidence, landslide, other), heavy rains/floods (washout/ scouring, flotation, mudslide, other), lightning (direct hit, secondary impact), temperature (thermal stress, frost heave frozen components, other), high winds, other natural force damage
External interference	Excavation damage (by operator, Operator's Contractor, third party, type of excavator, type of work performed), nearby industrial, man-made or other fire/explosion, damage by car, truck, or other motorized vehicle/equipment not engaged in excavation, damage by boats, barges, drilling rigs or other maritime equipment, previous mechanical damage not related to excavation, intentional damage (vandalism, terrorism, theft of transported commodity or equipment, other), other outside force damage
Construction /material defect	Material (body of pipe, component, joint), weld (butt, filet, pipe seam), malfunction of control/ relief equipment, threaded connection/ coupling failure, non-threaded connection failure, pump or pump-related equipment
	Other information

A.2.1. Accidents included in analysis

Accident data[2] are presented in the official website PHMSA [12]. Primarily the following objects have been selected among all data – onshore cross-country

---

[2] In this annex for convenience, an accident is defined as any case of accidental (uncontrolled) leak of OP, even if it may be attributed to incidents by one or other characteristics under the Federal Law No. 116-FZ

pipelines that are designed for crude oil transportation. The accidents were selected that occurred only in the line pipe, including valve sites.

The fact of change in accidents inclusion criteria in PHMSA database (Appendix 1, Table A.1-4) deserves special attention. Since 2002, the minimum spill volume, which operating companies are required to report, was reduced from 8 m^3 (50 barrels) up to 0.019 m^3 (0.12 barrels or 19 liters), which in turn significantly affected the number of accidents recorded in the database for this period: 946 accidents were recorded for the period 1984–2001 and 905 – for the period 2002–2009 (Fig. A.2-1). Moreover, in the period 2002–2009 the revision "rev 01-2001" of the form *PHMSA F 7000-1* of incident report was in force. According to instructions to this revision in case of so-called small spills over 0.019 m^3 (0.12 barrels / 19 liters) but less than 0.8 m^3 (5 barrels) only basic information was indicated: name of organization and contact details, time and place of incident, type of material, volumes of spilled and collected substance, cause of accidental leak, amount of damages caused to operator and third parties. This type of information is not representative. Since 2010, the form *PHMSA F 7000-1* again was subject to some changes related to provided accident data. Furthermore, one additional condition of inclusion in the accident database appeared: if spill volume is less than 5 barrels, and spill occurred as a result of a pipeline repair, such cases shall not be included in the database because they are not considered as accidents. Therefore, it was not noted such abrupt increase in the total number of accidents in 2010–2013, as it was in the previous period (Fig. A.2-1).

Based on the above stated information it can be concluded that inclusion in the database of accidents "with small spill volume" and "with small spill volumes in course of current repairs" was the cause of abrupt increase in the number of accidents in 2002–2009. Such accidents are easily distinguished from the others, as their description contains severely limited information (so-called "blank").

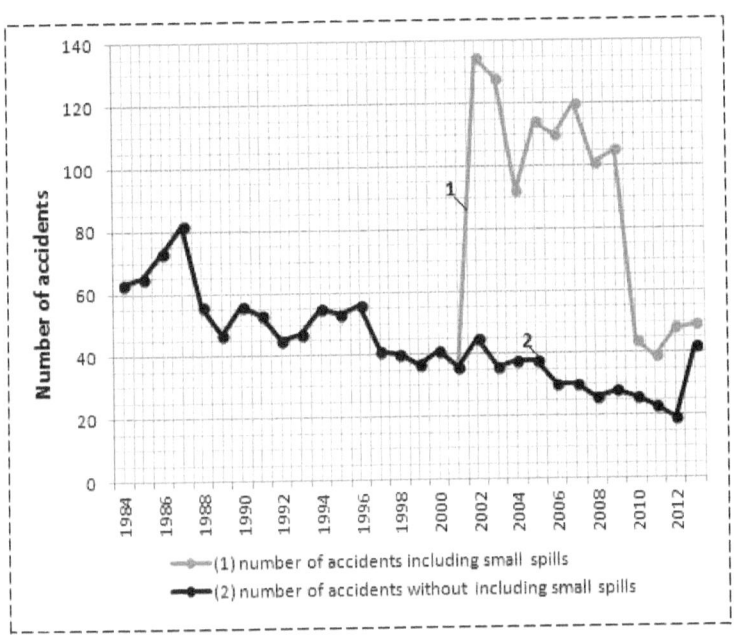

Figure A.2-1 – Number of accidents including/ without including small spills in 2002-2013

All documented cases of non-accidental leak of OP should be excluded from the OP accidental leak frequency calculation. It can be assumed that the cause of "low-volume spill" may be any of six classes of causes given in Table A.1-9 of Appendix 1, and of accidents "with small spill volumes in course of current repairs" – "Incorrect operation" and partly "Construction /material defect". According to the description of cause classes (Appendix 1, Table A.1-9), violation of technical regulations requirements (including for maintenance and repair), as well as personnel error related to incorrectly installed equipment are referred only to "Incorrect operation". Table A.2-2 presents data on the number of accidents by the cause class "Incorrect operation" for the period 2002–2013.

Given the data in Table A.2-2, as well as the fact that in the period 1984–2001 the number of accidents per year by the cause class "Incorrect operation" ranges from 0 to 3, it can be assumed that all so-called "blank" accidents recorded in 2002–2009 are accidents "with small spill volumes in course of current repairs", so they were excluded from the total number of accidents "with small spill volumes" for the period 2002–2013.

Table A.2-2 – Accidents by the cause class "Incorrect operation" for the period 2002–2013

Year	Total number of accidents in the cause class "Incorrect operation"	Number of "blank" accidents	Number of documented accidents
2002	10	10	0
2003	10	10	0
2004	3	2	1
2005	9	8	1
2006	10	9	1
2007	6	4	2
2008	9	9	0
2009	6	5	1
2010	4	3	4
2011	1	0	1
2012	2	2	2
2013	3	1	3
Total:	73	63	16

Note: Cells with the number of accidents taken into account in the further analysis are marked in bold font in Table A.2-2

One of the main causes of accidents "with small spill volumes" may be "Construction /material defect" (Appendix 1, Table A.1-9), but is unlikely that such accidents may occur in course of repair. Nevertheless, comparison of the number of accidents of the cause class "Construction /material defect" in 2002–2009 and in 2010–2013. (Table A.2-3) means that accidents "with small spill volumes in course of current repairs" were also recorded for this cause, as the number of accidents during specified periods differs significantly (more than twice!). Most likely, the accident rate "with small spill volumes" in the class of causes "Construction /material defect" at different years will be about the same. Therefore, it can be assumed that the total number of accidents in the cause class "Construction /material defect" in 2002–2009 can be reduced 2.5 times (Table A.2-3). This will be considered as exception of accidental leak cases "with small spill volumes in course of current repair" in the cause class "Construction /material defect".

Table A.2-3 – Accidents in the cause class "Construction /material defect" in 2002–2013

Год	Total number of accidents in the cause class "Construction /material defect"	Number of "blank" accidents	Number, reduced by 2.5 times	Number of "blank" accidents	Number of accidents with full description
2002	51	43	*20*	12	8
2003	44	41	*18*	15	3
2004	34	30	*14*	10	4
2005	46	36	*18*	8	10
2006	39	36	*16*	13	3
2007	55	51	*22*	18	4
2008	46	37	*18*	9	9
2009	49	42	*20*	13	7
2010	*17*	11	-	-	6
2011	*8*	5	-	-	3
2012	7	5	-	-	2
2013	*9*	1	-	-	8
Total:	405	338	*187*	120	67

Note: Cells with the number of accidents, corresponding to average level of accidents in the cause class of "Construction /material defect", which are accounted in the further analysis, are marked in bold italic font in Table A.2-3, in bold font – also accounted in the further analysis

The database also contains accidents in the period 2010–2013 with limited description ("blank"), but all they are referred only to accidents "with small spill volumes".

Therefore, in the future for accidental leak frequency of OP calculation (including small leaks), accidents were selected that occurred in the period 2004–2013, including accidents "with small spill volumes" (total 623 accidents). For the database detailed analysis performance in order to develop the procedure for expected accidental leak frequency calculation of each OP section, the accidents were selected that occurred in the period 1984–2013 except accidents "with small spill volumes" and "with small spill volumes in course of current repair" in the period 2002–2013 (total of 1,327 accidents) (Figure A.2-1, black line – 2).

Distributions of selected accidents depending on the cause class of their occurrence are shown in Figures A.2-2–A.2-5.

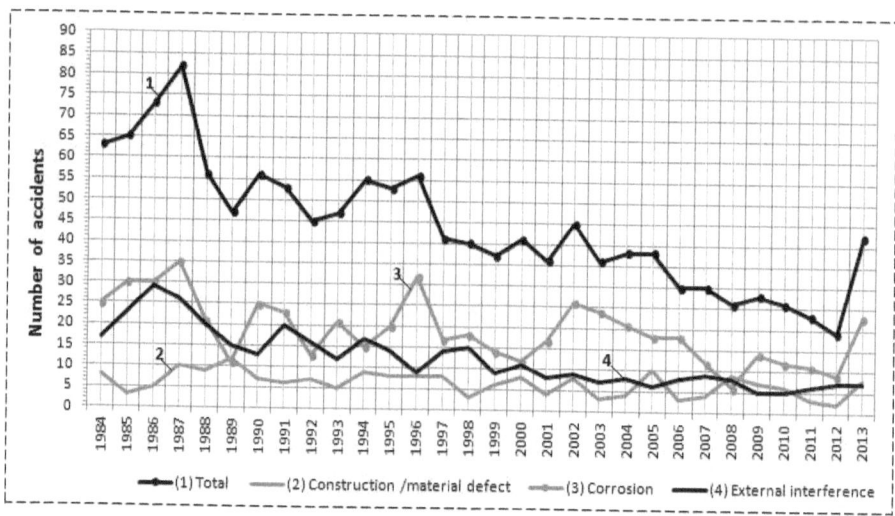

Figure A.2-2 – Number of accidents at OP (total and according to cause class) per each year of analyzed period (1984–2013)

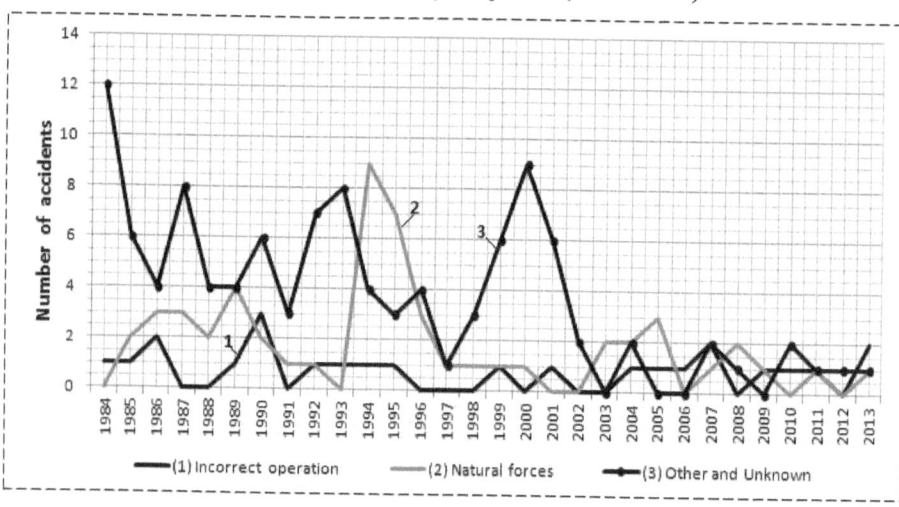

Figure A.2-3 – Number of accidents at OP (according to cause class) per each year of analyzed period (1984–2013)

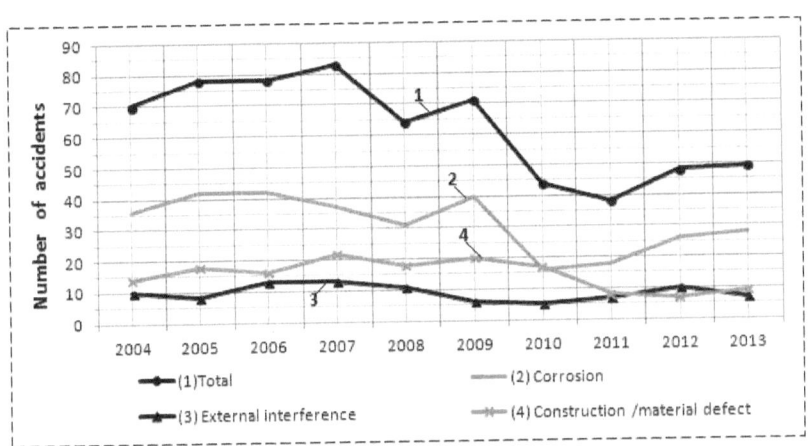

Figure A.2-4 – Number of accidents at OP (total and according to cause class) per each year of analyzed period (2004–2013) (including small spills)

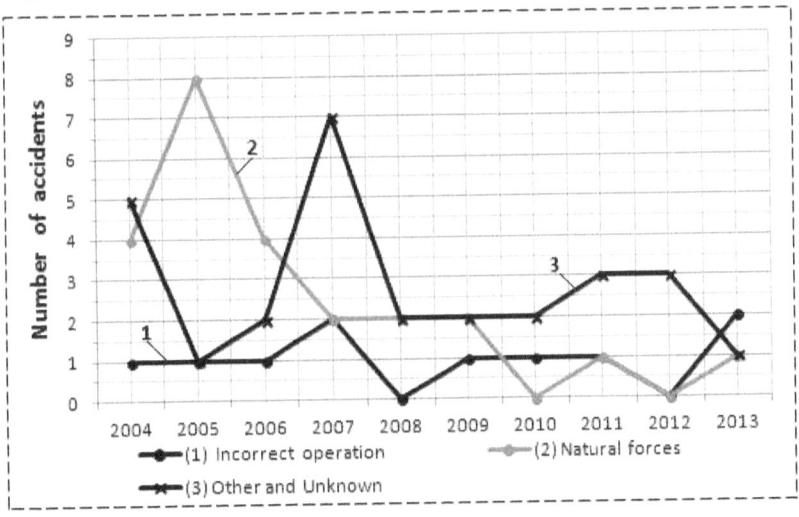

Figure A.2-5 – Number of accidents at OP (according to cause class) per each year of analyzed period (2004–2013) (including small spills)

Tables A.2-4–A.2-6 show the distribution of selected accidents depending on the cause class of their occurrence in the periods 1984–2013 and 2004–2013.

It seems appropriate to note also that almost 80 % of selected 1,327 accidents in the period 1984–2013 occurred directly on the pipe "body", and the rest of accidents took place in the welds, stop valves, bolted connections, welded fittings etc.

Table A.2-4 – Total number of accidents according to cause class (1984-2013)

Cause class accidents	Number of accidents	Overall percentage, %
Construction /material defect	193	14,54
Incorrect operation	24	1,81
Corrosion	573	43,18
Natural forces	54	4,07
External interference	373	28,11
Other and Unknown	110	8,29
Total:	**1327**	**100,00**

Table A.2-5 – Total number of accidents according to cause class (2004-2013)

Cause class accidents	Number of accidents	Overall percentage, %
Construction /material defect	56	18,67
Incorrect operation	10	3,33
Corrosion	144	48,00
Natural forces	11	3,67
External interference	69	23,00
Other and Unknown	10	3,33
Total:	**300**	**100,00**

Table A.2-6 – Total number of accidents according to cause class (2004-2013)

(including small spills)

Cause class accidents	Number of accidents	Overall percentage, %
Construction /material defect [3]	149	23,92
Incorrect operation	16	2,57
Corrosion	316	50,72
Natural forces	24	3,85
External interference	90	14,45
Other and Unknown	28	4,49
Total:	**623**	**100,00**

---

[3] The abovementioned assumption about exclusion of accidents "with small spill volumes in course of current repair" could somewhat lower the accident rate on cause class "Construction /material defect" but generally this effect is inconsequential.

A.2.2. OP exposure

The accidental leak frequency of OP is calculated by dividing the number of accidents by the exposure. The exposure means the length of a pipeline supervised within some period, measured in [kilometers–years]. It can be defined both as the total exposure of the whole transport system, and as the part of system – e.g. piping exposure of certain diameter or of specified depth of cover. Also the exposure period calculation not necessarily shall be taken to be equal to one year; usually sufficiently longer intervals are taken – five, ten, twenty years, etc.

It is obvious that the bigger is the exposure of pipelines taken to the statistical analysis, the more accurate is the resulting accidental leak frequency of OP. The exposure is determined on the basis of annual reports on the state of supervised pipeline system (the form *PHMSA F 7000-1.1*) [11, 13] (Table A.2-7).

Table A.2-7 – Supervised OP information

Reported information:	
Operator	Name and address
Commodity	Commodity group (crude oil, refined and/or petroleum product, highly volatile liquids, carbon dioxide, fuel grade ethanol), volume transported in barrel-miles
Miles of pipe	By operating pipelines; by onshore/ offshore in accordance with nominal pipe size (diameter), decade installed, corrosion prevention status; by onshore/ offshore in accordance with electric resistance welded (ERW) by weld type and decade; by onshore/ offshore in accordance with minimum yield strength; total segment miles that could affect high consequences areas (HCAs)
	And other

Operating companies started providing this kind of reports only since 2004. This makes the PHMSA database less representative and "sensitive" to changes in the basic settings of OP (such as: diameter, wall thickness, depth of cover, etc.) that affect the operating safety. The EGIG database [4] of GP may serve as the "Reference" database on accidents. The main problem is that formation of accidents databases of OP mainly is aimed at fixing spilled and collected volumes of oil and oil products, the area of contamination, the amount of damage caused by accidents. When forming the accident database of GP, EGIG experts pay great attention to fixation of GP characteristics, where the accident have occurred (diameter, wall thickness, depth of cover, etc.). Moreover, very detailed information is collected on all accountable GPs, which allows determining the GP exposure depending on various parameters. This approach to statistical data collection enables the effectiveness assessment performing of sufficiently wide range of applied GP safety measures, based on results of quantitative risk assessment, in other words (simplistically) – based on expected accidental frequency calculation of GP. The difference in the approach to statistical data collection of OP and GP is caused, in the first place, by difference in the specifics of an accident occurrence and development. GP accidents develop quickly enough and may result in large number of victims unlike OP accidents. It is impossible to say that consequences of OP accidents are less extensive, but the comparative analysis of GP [3, 10] and OP (Appendix 1) accident databases shows the significant difference in the principles of statistical data collection. More simply it can be said, that experts of EGIG [4] group are more focused on the fact that the statistical database allows to "feel" the difference in the application of various methods of preventing the accidental leak of GP, whereas CONCAWE [6], NEB [7], PHMSA [9] databases and other organizations are more focused on possible consequences of OP accidents. The difference of PHMSA database is that collected and systematized data on the state of operated OP, starting since 2004, still allow tracking of regularities in the use of various safety measures. It will require making a number of assumptions described below. It is also important the fact that PHMSA started since 1970 collecting such kind of data related to cross-country, distribution and gathering pipelines.

Based on the summary data analysis of all supervised pipelines for the period 2004–2013, it was found that the average length of onshore cross-country pipelines, transporting just crude oil, was about 27 % of the total length of all pipelines included in the PHMSA base. The general data on the length of all pipelines (regardless of the transported material) included in the PHMSA database [13] are known for the period since 1984.

Therefore, further calculation of the total exposure of all onshore OP for the full reporting period 1984–2013 (Fig. A.2-6) was performed, taking into account the above average share of onshore OP, and at the end of 2013 it was 2,124.4 thousand km per year.

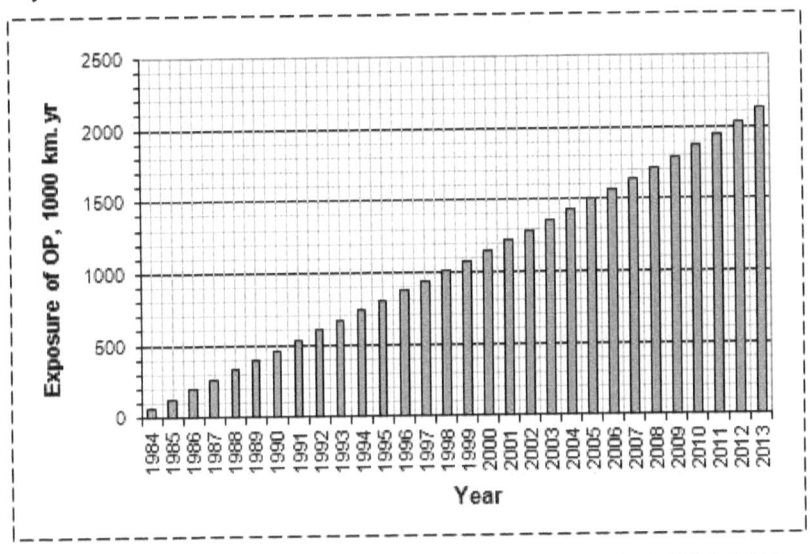

Figure A.2-6 – Exposure of onshore OP for the period 1984–2013

Similarly, the exposure of onshore OP was identified depending on diameter (Fig. A.2-7, A.2-8), taking into account the average share of pipelines with a particular diameter (Table A.2-8), calculated on the basis of reported data for the period 2004–2013.

Table A.2-8 – Average share of OP with a particular diameter of total number of onshore OP included in PHMSA database in 2004–2013

Pipe diameter, "	Average share, %	Pipe diameter, "	Average share, %
4[4] or less	1,40	22	5,20
6	5,43	24	6,50
8	14,96	26	2,16
10	14,13	28	0,03
12	11,11	30	4,45
14	1,81	32	0,01
16	13,08	34	1,67
18	2,37	36	1,06
20	10,54	36 and over	4,09

---

[4] In further analysis "4 or less" combination was substituted with "0-4".

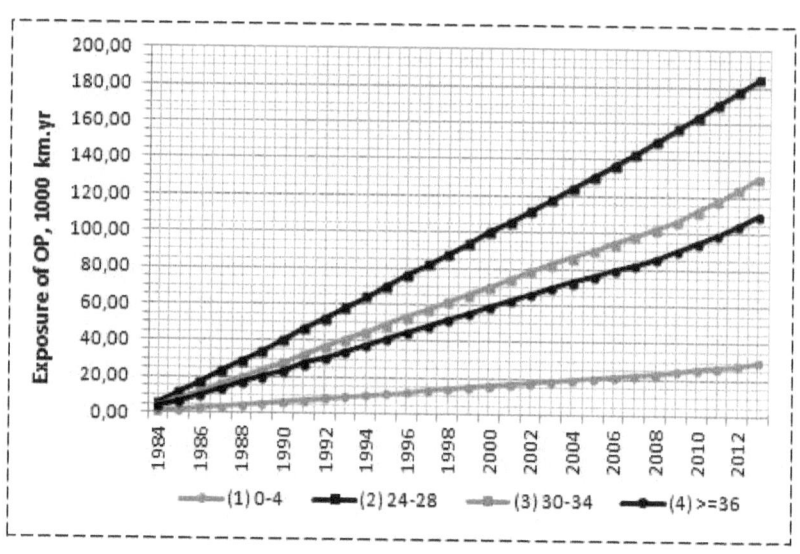

Figure A.2-7 – Exposure of OP for the period 1984–2013 depending on diameter,"

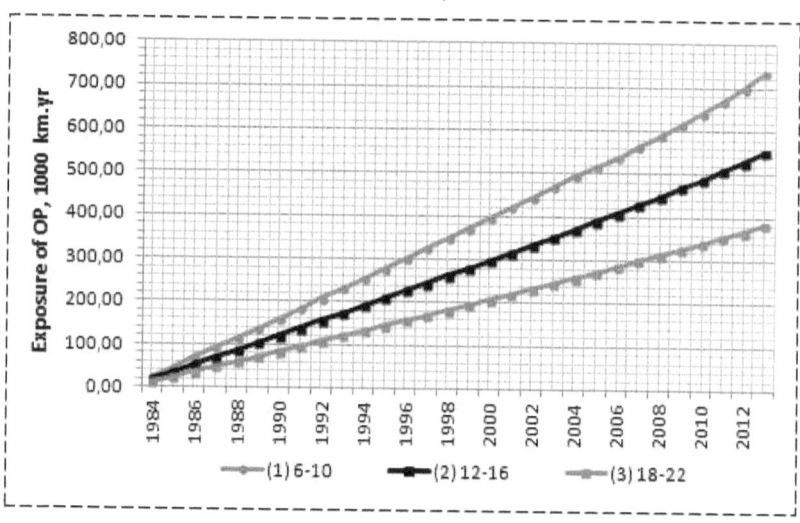

Figure A.2-8 – Exposure of OP for the period 1984–2013 depending on diameter,"

Taking into account the average share of onshore OP depending on the presence/absence of cathodic protection and/or corrosion-resistant coating, obtained on the basis of reported data in 2004–2013, the corresponding exposure was assigned at the end of 2013 (Table A.2-9).

Table A.2-9 – Average share of the total number of onshore OP included in PHMSA database (2004–2013), and exposure of OP (1984–2013) depending on the presence / absence of cathodic protection and/or corrosion-resistant coating

Corrosion protection	Average share, % (2004-2013)	Exposure, 1000 km.yr (1984-2013 гг.)
Cathodically protected:		
Bare	2,94	62,22 (Figure A.2-9)
Coated	95,37	2025,27 (Figure A.2-10)
Cathodically unprotected:		
Bare	0,59	12,29 (Figure A.2-9)
Coated	1,10	23,38 (Figure A.2-9)
	100	

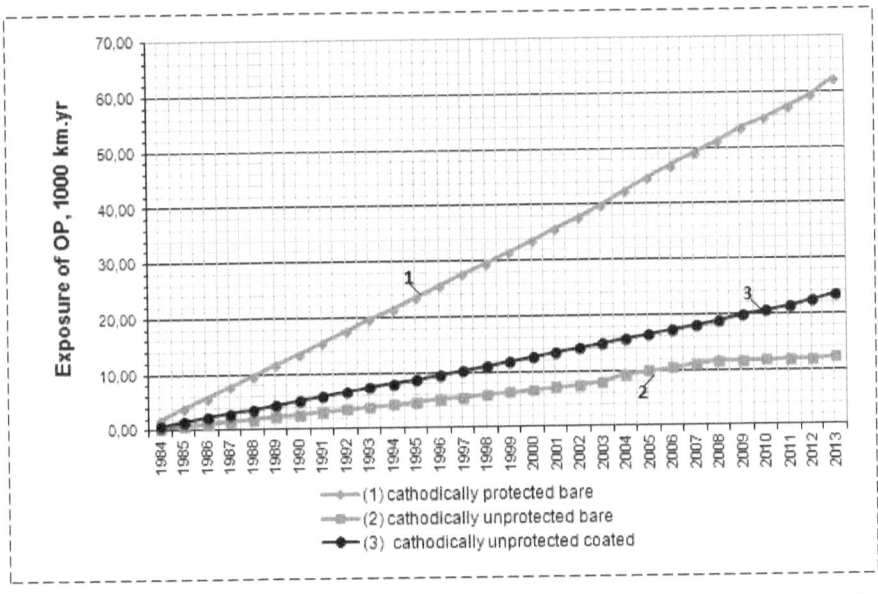

Figure A.2-9 – Exposure of OP depending on material and corrosion prevention status for the period 1984–2013

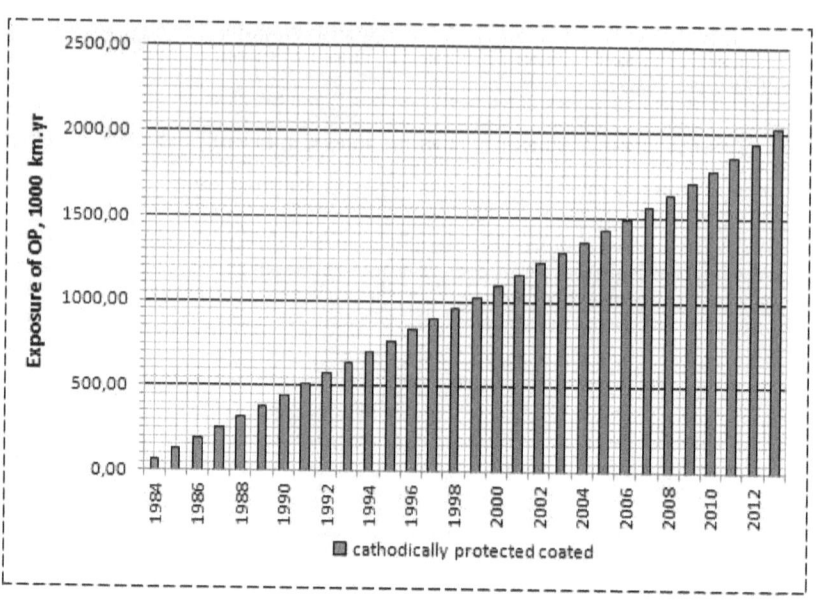

Figure A.2-10 – Exposure of cathodically protected coated OP for the period 1984–2014

Fig. A.2-11–A.2-13 present exposure of onshore OP in 2004–2013 per decade installed.

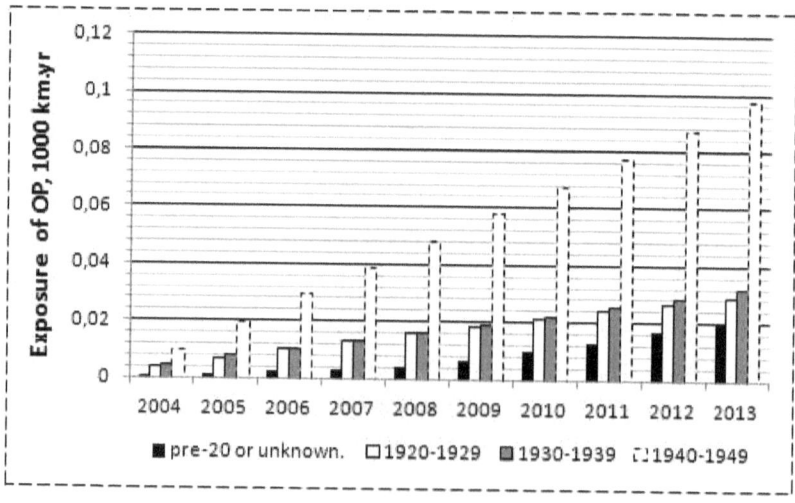

Figure A.2-11 – Exposure of OP per decade installed (pre-1920 or unknown, 1920-1929, 1930-1939 and 1940-1949)

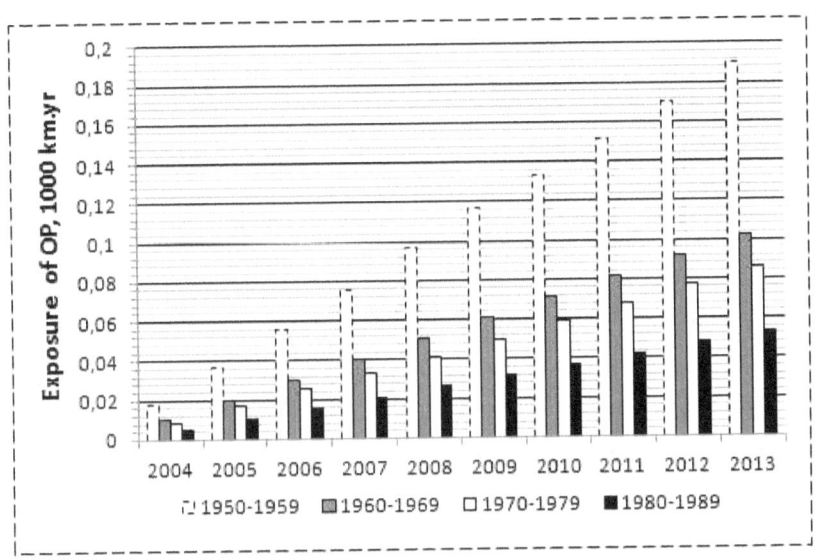

Figure A.2-12 – Exposure of OP per decade installed (1950-1959, 1960-1969, 1970-1979 and 1980-1989)

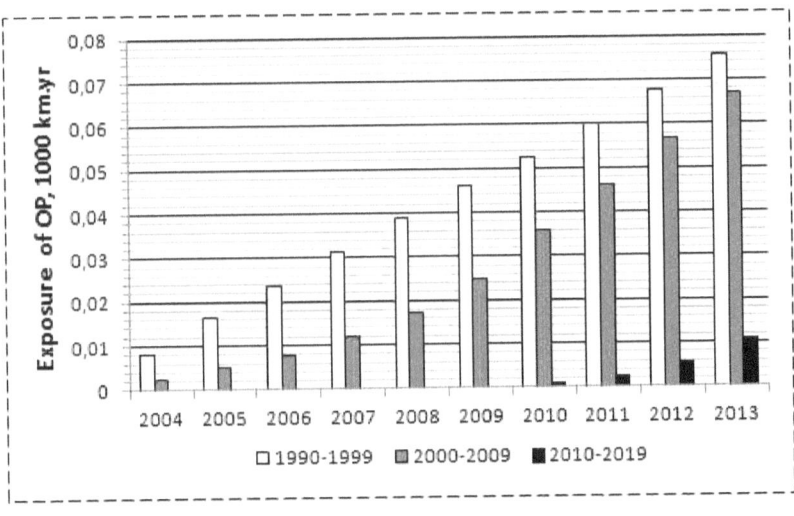

Figure A.2-13 – Exposure of OP per decade installed (1990-1999, 2000-2009 and 2010-2019)

Unfortunately, the aggregate collected PHMSA data on the state of pipeline system and on accidents do not allow considering the gathering lines separately. According to the instructions on filling out annual reports on the state of supervised pipeline system (form *PHMSA F 7000-1.1*) [11], operating companies indicate only the general (offshore and onshore) length of gathering lines that include pipelines with the nominal outside diameter 219.1 mm (8–5/8") or less, transporting crude oil from production facilities. However, it seems impossible to identify these pipelines only by diameter because, for example, in 2004, the total length of gathering lines was 3,533.69 km, while the total length of onshore pipelines with the diameter less than 8" – 14,058.82 km, offshore – 1,904.66 km. Therefore, it seems inappropriate to exclude from further consideration all the pipelines with the diameter less than 8" (and hence accidents registered at them), since most of them for unknown to Author reasons have not been referred to gathering lines, despite the relatively small diameter.

In addition, PHMSA data [13] do not allow performing separate analysis of accidents in underground and aboveground OP, since it is impossible to determine the appropriate exposure.

APPENDIX 3.
PROCEDURE DEVELOPMENT FOR ACCIDENTAL LEAK FREQUENCY CALCULATION

The logic diagram of OP accident condition may be represented as a "fault tree". Based on the experience of accidents causes classification in the European and American oil pipelines (Appendix 1, Table A.1-6), the number of accidents cause classes was assumed to be equal to six (Fig. 1).

The estimated accident frequency F in any OP section is calculated by the formula:

$$F = \sum_{i=1}^{S} f_i \qquad (A.3-1)$$

where f_i – accidental leak frequency for the *i-th* cause class (1/1,000 km·yr);
S – number of classes of leak causes (where S = 6).

Proposed methodology of OP accidental leak frequency calculation is analogous to the procedure of GP accidental leak frequency calculation presented in [3, 5].

A.3.1 Distribution of accidental leak frequency of OP depending on cause class

Distribution of accidental leak frequency of OP was calculated for each accidents cause class (based on the data in Appendix 2).

Distributions presented in Table A.3-1 refer to the entire period of observation and data collection 1984–2013 (total of 1,327 accidents occurred, the OP exposure is 2,124.4 thous. km per year) (Appendix 2, Table A.2-4, Figure A.2-6).

Accidental leak frequencies of OP were also calculated for the period 2004–2013. (total of 300 accidents occurred, OP exposure is 767.87 thous. km per year) (Appendix 2, Table A.2-5) that are presented in Table A.3-2.

Table A.3-3 presents accidental leak frequencies of OP calculated for the period 2004–2013, taking into account accidents "with small spills" (total of 623 accidents occurred, OP exposure is 767.87 thous. km per year) (Appendix 2, Table A.2-6).

Table A.3-1 – Primary accidental leak frequency of OP depending on cause class (for the whole period of supervision 1984–2013)

Cause class	Primary accidental leak frequency (1/1000 km.yr)	Overall percentage, %
Construction /material defect	$9,1 \cdot 10^{-2}$	14,54%
Incorrect operation	$1,1 \cdot 10^{-2}$	1,81%
Corrosion	$2,7 \cdot 10^{-1}$	43,18%
Natural forces	$2,5 \cdot 10^{-2}$	4,07%
External interference	$1,76 \cdot 10^{-1}$	28,11%
Other and Unknown	$5,2 \cdot 10^{-2}$	8,29%
Total:	$6,25 \cdot 10^{-1}$	100%

Table A.3-2 – Primary accidental leak frequency of OP depending on cause class (2004-2013)

Cause class	Primary accidental leak frequency (1/1000 km.yr)	Overall percentage, %
Construction /material defect	$7,3 \cdot 10^{-2}$	18,67%
Incorrect operation	$1,3 \cdot 10^{-2}$	3,33%
Corrosion	$1,88 \cdot 10^{-1}$	48,00%
Natural forces	$1,4 \cdot 10^{-2}$	3,67%
External interference	$9,0 \cdot 10^{-2}$	23,00%
Other and Unknown	$1,3 \cdot 10^{-2}$	3,33%
Total:	$3,91 \cdot 10^{-1}$	100%

Table A.3-3 – Primary accidental leak frequency of OP depending on cause class (2004-2013, including small spills)

Cause class	Primary accidental leak frequency (1/1000 km.yr)	Overall percentage, %
Construction /material defect	$1,94 \cdot 10^{-1}$	23,92%
Incorrect operation	$2,1 \cdot 10^{-2}$	2,57%
Corrosion	$4,12 \cdot 10^{-1}$	50,72%
Natural forces	$3,1 \cdot 10^{-2}$	3,85%
External interference	$1,17 \cdot 10^{-1}$	14,45%
Other and Unknown	$3,6 \cdot 10^{-2}$	4,49%
Total:	$8,11 \cdot 10^{-1}$	100%

The values of frequencies in Table A.3-2 can be used as primary (average) in the accidental leak frequency of OP calculation, given that in the future measures will be planned for oil spill response, as oil spill incidents included in the statistics for of these frequencies calculation mainly are accidental (Appendix 2, Part A.2.1). The values of frequency in Table A.3-3 can be taken as primary (average) in the calculation of expected accidental leak frequency of OP, if the purpose of calculations is to determine the expected level of OP accidental leak, taking into account not only oil spills, but also possible smaller leaks (incidents), since as was noted in (Appendix 2, Part A.2.1), accidents selected for calculation include incidents from 0.019 m^3. This kind of OP accidents occurs much more frequently, so for the period 2004–2013 the total value of frequency increased in comparison with values presented in Table A.3-2. Calculated primary frequencies present values, which are statistically averaged over all OP included in the statistics of PHMSA [9], i.e. corresponding to some "average" OP [3]. Procedures for these data use to calculate expected accidental leak frequencies with the account of specific characteristics of OP, as well as natural features of the route will be given below.

A.3.2 Distribution of accidental leak frequency of OP per damage size

The data on types of accident holes started being included in PHMSA database [9] only since 2002. The following types of damage are distinguished in PHMSA [12]: pinhole; puncture; tear / crack (longitudinal); circumferential separation or, in other words, full rupture (100 % of cross-section) (hereinafter Rupture); connection failure (e.g. in the shut-off valves, etc.). The data on accidents for the period 2002–2013 were included in the analysis. Results are shown in Table A.3-4.

According to Table A.3-4, for the period 2002–2013 the type of damage is known for 79.00 % accidents; 11.81 % – the type of damage is unknown for accidents cause classes: "Construction /material defect", "Corrosion", "External interference" and "Natural forces"; the remaining 9.19 % refer to "Incorrect operation" and "Other and unknown" causes without information about types of damage (noted as "unknown"). Accident causes classes "Incorrect operation" and "Other and unknown" unlikely may cause large number of accidents with "Rupture".

In accidents of the rest cause classes with initially not specified type of damage (11.81 %) it is impossible to exclude completely the occurrence of "Rupture".

Therefore, the total amount of specified "unknown" 21.00 % accidents was conservatively distributed equally among the groups "Leaks" and "Rupture". This distribution resulted in the following ratio: 87.14 % of the total number of accidents are accidents with the type of damage related to the group "Leaks" and 12.86 % of the total number of accidents are accidents with the type of damage "Rupture".

Therefore, the accidental leak frequency of each OP section may be divided for further calculations into two specified groups according to this ratio (overall frequency 0.42 1/1,000 km·yr divided into 0.366 and 0.054 respectively).

It is necessary to know the scale of pipeline damage for accident consequences calculation, i.e. the size of accident hole. Unfortunately, PHMSA database [9, 12] practically does not contain information on damage sizes (of 381 considered accidents, sizes are presented only for 68). But, nevertheless, these types of damage may be distributed by holes sizes (e.g., holes with average diameter (mm) 12.5; 25; 50; 100; > 150 – rupture), using expert assessment method as it was done for GP [5] on the basis of data of EGIG (European Gas Pipeline Incident Data Group).

Table A.3-4 – Damage types depending on accidental leak cause class of OP (2002–2013)

Cause	Leaks:						
	Pinhole	Puncture	Tear/crack	Connection failure	Rupture	Unknown	Overall
Construction /material defect (frequency, 1/1000 km.yr)	0,014	0,000	0,020	0,018	0,003	0,019	0,074
Frequency per cause, %	19,40	0,00	26,87	23,88	4,48	25,37	100,00
Overall frequency, %	3,41	0,00	4,72	4,20	0,79	4,46	17,59
Corrosion (frequency, 1/1000 km.yr)	0,198	0,000	0,003	0,001	0,001	0,010	0,214
Frequency per cause, %	92,78	0,00	1,55	0,52	0,52	4,64	100,00
Overall frequency, %	47,24	0,00	0,79	0,26	0,26	2,36	50,92
External interference (frequency, 1/1000 km.yr)	0,004	0,063	0,008	0,001	0,001	0,017	0,094
Frequency per cause, %	4,71	67,06	8,24	1,18	1,18	17,65	100,00
Overall frequency, %	1,05	14,96	1,84	0,26	0,26	3,94	22,31
Natural forces (frequency, 1/1000 km.yr)	0,004	0,000	0,000	0,001	0,004	0,004	0,014
Frequency per cause, %	30,77	0,00	0,00	7,69	30,77	30,77	100,00
Overall frequency, %	1,05	0,00	0,00	0,26	1,05	1,05	3,41
Incorrect operation (frequency, 1/1000 km.yr)	n/a	n/a	n/a	n/a	n/a	n/a	0,011
Frequency per cause, %	-	-	-	-	-	-	100,00
Overall frequency, %	-	-	-	-	-	-	2,62
Other and Unknown (frequency, 1/1000 km.yr)	n/a	n/a	n/a	n/a	n/a	n/a	0,013
Frequency per cause, %	-	-	-	-	-	-	

Cause	Leaks:				Rupture	Unknown	Overall
	Pinhole	Puncture	Tear/crack	Connection failure			
Overall frequency, %	-	-	-	-	-	-	3,15
Total percentage per overall frequency, %	49,34	14,96	7,35	4,99	2,36	11,81	0,420
Not available data, %	11,81+9,19					-	-
Grand total (with account of 21 % N/A data repartitioning), %	87,14				12,86	-	-
Overall frequency, (1/1000 km.yr)	0,366				0,054	-	-

A.3.3 Parameters influencing the accidental leak frequency of OP

Parameters presented in [3] for GP, based on PHMSA statistical data [9] in a varying degree can be taken into account in the OP accidental leak frequency calculation (Table A.3-5).

In general, the resulting accidental leak frequency for each OP section is calculated by summing six components that corresponds to accidental leak frequencies of this OP by each of six accidental leak cause classes according to the formula:

$$F_k(m) = \sum_{i=1}^{S} f_{ik}(m) \qquad (A.3\text{-}2)$$

where $F_k(m)$ – accidental leak frequency for the k-th damage size of the m-th section of OP (1/1,000 km·yr);

$f_{ik}(m)$ – accidental leak frequency for the k-th damage size of the i-th class of accidental leak causes of the m-th section of OP (1/1,000 km·yr);

$i = \{1\ldots6\}$ – classes of accidental leak causes of OP (number of classes – $S=6$);

$k = \{1\ldots2\}$ or $\{1\ldots5\}$ – types (sizes) of damages: leak / full rupture (100 % cross section) $\{1\ldots2\}$ (Table A.3-4) or diameters (mm) of equivalent emergency openings: 12.5; 25; 50; 100; > 150 – rupture $\{1 \ldots 5\}$ (Part A.3.2).

Table A.3-5 – Summary table of parameters taken into account in accidental leak frequency calculation of OP

Cause Parameters	External interference	Construction / material defect	Corrosion	Natural forces	Incorrect operation	Other and Unknown
Pipe diameter	-	-	-	+	+	-
Wall thickness	+	-	+	+	-	-
Coating type	-	-	+	-	-	-
Depth of Cover	+	-	-	+	-	-
HDD [5]	+	-	-	+	-	-

---

[5] OP laying by horizontal directional drilling (HDD) method.

Cause Parameters	External interference	Construction / material defect	Corrosion	Natural forces	Incorrect operation	Other and Unknown
Railways, roads, utilities crossings	+	-	-	-	-	-
Seismic activity, Tectonic fault	-	-	-	+	-	-
Geo hazards location crossings	-	-	-	+	-	-

Below is given description of procedures for accidental leak frequency calculation of OP at specified section for each of six cause classes, based on PHMSA statistical data [9].

A.3.4 Accidental leak cause class: "External interference"

According to the performed analysis of PHMSA database [12], "External interference" is the second largest cause class of leaks over the entire observation period 1984–2013. As shown in Table A.3-1, accidental leak frequency for this cause class for the period 1984–2013 is 0.176 cases per 1,000 km·yr.

Pipe diameter

Figure A.3-1 shows that the frequency of accidents by the cause class "External interference" increases with pipeline diameter decreasing. This is not necessarily directly related to the pipeline diameter and can be subject of wall thickness dependence on the pipe diameter, as pipes of large diameter generally have bigger wall thickness [3]. As shown in Figure A.3-1, pipelines of large diameter (42" and more) are practically not exposed to risk of accidents due to external anthropogenic impacts.

Shown in Figure A.3-1 dependence of accidental leak frequency by the cause class "External interference" on the pipe diameter can be numerically presented in the form of regression:

$$F_{EI}=0,6951 \cdot exp(-0,1 \cdot D), \qquad (A.3-3)$$

где f_{EI} – OP accidental leak frequency by the cause class "External interference";
D – OP diameter, ".

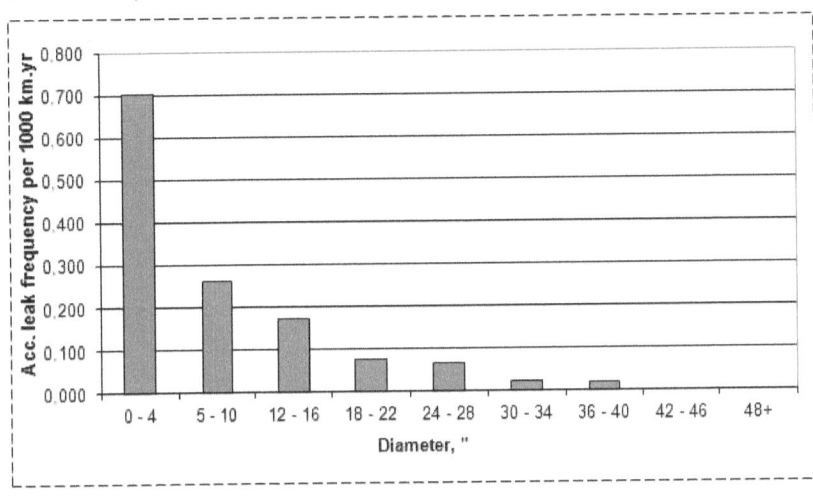

Figure A.3-1 – Relationship between OP diameter and accidental leak frequency due to "External interference" (1984–2013)

Therefore, Table A.3-6 shows calculation results of accidental leak frequency reduction factors by the cause class "External interference" for OP different diameters (k_D) with account of dependence A.3-3.

Table A.3-6 k_p values dependence on OP diameter

Diameter, "	Primary accidental leak frequency, 1000 km.yr	Accidental leak frequency per dependence A.3-3, 1000 km.yr	k_D
48		0,0057	0,03
40		0,0127	0,07
30		0,0346	0,2
20	0,176	0,0941	0,54
14		0,1714	0,98
8		0,3123	1,78
4		0,4659	2,65

Pipe wall thickness

PHMSA data (Appendix 2, part A.2.2, [13]) do not allow to determine separately the OP exposure per wall thickness, therefore the mathematical relationship between accidental leak frequency of OP and wall thickness cannot be determined. Table A.3-7 shows data on the number of accidents distribution due to the cause "External interference" depending on the pipe wall thickness; according to these data it can be stated that OP with wall thickness more than 10 mm actually are not subjected to accidents due to this cause.

Table A.3-7 – Number of accidents due to the cause class "External interference" per pipe wall thickness (1984–2013)

Pipe wall thickness, mm	0-5	5-10	10-15	>15
Number of accidents due to the cause class "External interference"	36	307	1	3

Proportion of OP accidents with wall thickness more than 10 mm of the total amount (373 accidents) (Appendix 2, Table A.2-4) is 0.011, whereas OP with a wall thickness less than 10 mm–0.92 (0.07 – proportion of accidents with unknown wall thickness). The similar ratio can be used for frequency reduction factor assignment. However, since it is unknown the ratio of lengths of all observed OP with wall thickness up to 10 mm and above 10 mm to the total exposure, it seems appropriate to assign slightly larger (twice as much) frequency reduction factor than the above ratio. It will allow compensation of possible differences in exposure depending on OP wall thickness.

Based on the above, the accidental leak frequency reduction factor of OP (k_{WT}) with wall thickness more than 10 mm may be taken equal to 0.022. For OP with wall thickness less than 10 mm the frequency reduction factor is taken equal to 1[6].

It is necessary to take into account also the fact that with pipeline diameter increasing in most cases the wall thickness increases as well. Therefore, to avoid "double counting" of influencing factors, the index may be used that accounts the

---

[6] Since 92 % of accidents, accounted for OP accidental leak primary frequency calculation by the cause class "External interference", took place at OP with wall thickness less than 10 mm, it seems appropriate do not perform frequency correction of such pipelines, therefore $k_{WT} = 1$.

influence of wall thickness only in calculations for the same pipeline diameter. In this case, it can be assumed conservatively that the accidental leak frequency by the cause class "External interference" does not depend on pipeline diameter, and only one factor, i.e. k_{WT}, shall be applied in calculations.

Pipe depth of cover

PHMSA data (Appendix 2, part A.2.2, [13]) make impossible to determine the OP exposure separately on depth of cover; furthermore, information on the depth of cover at accidents is available in the database only for periods 1984–1985 and 2002–2013 [12].

It should be noted that there is significant decrease in the number of accidents when OP depth of cover is in the range 0.8–1 m, with consequent increase in the number of accidents, when OP depth of cover exceeds 1 m (Table A.3-8). It can be assumed that the depth of cover in the range 0.8–1 m is not considered as significant (i.e. safe "by default"), and thereby all possible measures are taken for visual indication of OP route and information of organizations and population in the area of OP location on the OP route, enhanced monitoring of the OP route, etc. When laying OP at the depth over 1 m, it is considered that the probability of external interference is negligible, perhaps it "weakens vigilance" of an operating organization and reduces the number of implemented measures to ensure safety of the route. Furthermore, this abrupt (twice) increasing of the number of OP accidents with depth of cover over 1 m may be associated with a significantly greater their total length in comparison with OP with depth of cover 0.8–1 m.

Table A.3-8 – Number of accidents due to the cause class "External interference" per depth of cover (1984-1985, 2002-2013)

Number of accidents due to the cause class "External interference"	Depth of cover, cm		
	0-80	80-100	>100
In a period of 1984-1985	24	2	4
In a period of 2002-2013	38	10	20

On the basis of available data, performance of reliable analysis of accidental leak frequency by the cause class "External interference" dependence on the depth of cover of OP is impossible, and the corresponding factors cannot be assigned.

Crossings performed by horizontal directional drilling

It can be assumed that in the areas of crossings performed by horizontal directional drilling (HDD), due to the large depth of cover of OP the input of external interference is completely excluded, the correction factor is equal to 0, while in other areas it is equal to 1 [3].

Railways, roads /utilities crossings

The cause of pipelines damage at roads, railways, utilities crossings consists in the possibility of external interference of third parties on OP (e.g. in the case of heavy construction machinery or excavation equipment use without approval of OP operating company). PHMSA data [12] is not enough for reliable analysis of dependence of accidental leak frequency by the cause class "External interference" on the OP route crossing with the above objects (this kind of information was included in the database only from 2010).

However, in the works [6, 14] it was noted that intensity of the frequency of OP damages in urban and rural areas is different. It seems logical, as in urban areas works on earth cover and roadway excavation are performed more frequently, and underground utilities are fairly dense. Therefore, the probability of pipeline damage caused by external interference is higher there. To calculate the frequency of OP damage [14] it is recommended to multiply the primary frequency by the factor 5 in urban areas, while in rural areas – by the factor 0.8. The values of these factors were obtained based on the European statistics CONCAWE (for the period 1971–1996). Given that in Russia OP usually do not pass directly through settlements, the pipeline systems are considerable longer and they can cross very large number of different types of roads, we assume for OP sections with road, railways, utilities crossings that accidental leak frequency caused by external interferences is 2 times higher than the frequency caused by the same cause class at the adjacent transition area. In this case, the estimated length of crossing through categorized roads and railways is recommended to be taken equal to the length of section 25 m on both sides of the edge elements of the road. The estimated length of other road crossings can be taken equal to 20 m, and the estimated length of the underground utilities crossing – 5 m [3].

Accidental leak frequency calculation of OP by the cause class "External interference"

Thereby, the accidental leak frequency of OP by the cause class "External interference" is calculated by the formula:

$$f_{EI} = f_{PEI} \cdot k_D \cdot k_{HDD} \cdot k_C, \quad (A.3-4)$$

where f_{EI} – OP accidental leak frequency by the cause class "External interference";

f_{PEI} – primary frequency of OP accidental leak by the cause class "External interference" (Table A.3-2 or A.3-3);

k_D – correction factor of OP accidental leak frequency by the cause class "External interference", taking into account the effect of OP diameter (taking into account the dependence A.3-3 or in Table A.3-6);

k_{HDD} – correction factor of OP accidental leak frequency by the cause class "External interference", taking into account the OP laying, by using the method of horizontal directional drilling: $k_{HDD} = 0$, if sections of OP crossings are performed using the HDD method; $k_{HDD} = 1$, for other OP sections;

k_C – correction factor of OP accidental leak frequency by the cause class "External interference", taking into account the effect of OP railways, roads and utilities crossings: $k_C=2$, if OP section has road, railways, facilities crossings; $k_C = 1$, for other OP sections.

To calculate accidental leak frequency of OP by the cause class "External interference" for a pipeline of the same diameter but with different wall thickness (i.e. the safety class) the following formula can be used:

$$f_{EI} = f_{PEI} \cdot k_{WT} \cdot k_{HDD} \cdot k_C, \qquad (A.3\text{-}5)$$

where k_{WT} – correction factor of OP accidental leak frequency by the cause class "External interference", taking into account the effect of OP wall thickness: k_{WT} =0,022 with OP wall thickness more than 10 mm; k_{WT} = 1, with OP wall thickness less than or equal to 10 mm.

A.3.5 Accidental leak cause class: "Construction /material defect"

As follows from Table A.3-1, the relative portion of accidents, which causes belong to the cause class "Construction /material defect", is 14.54 %, and the accidental leak frequency for the period 1984–2013 is equal to 0.091 cases per 1,000 km·yr.

Year of construction

Data on the level of accidents by the cause class "Construction /material defect" according to the year of construction of OP are presented in Figure A.3-2. Dependence is defined for the period 2004–2013.

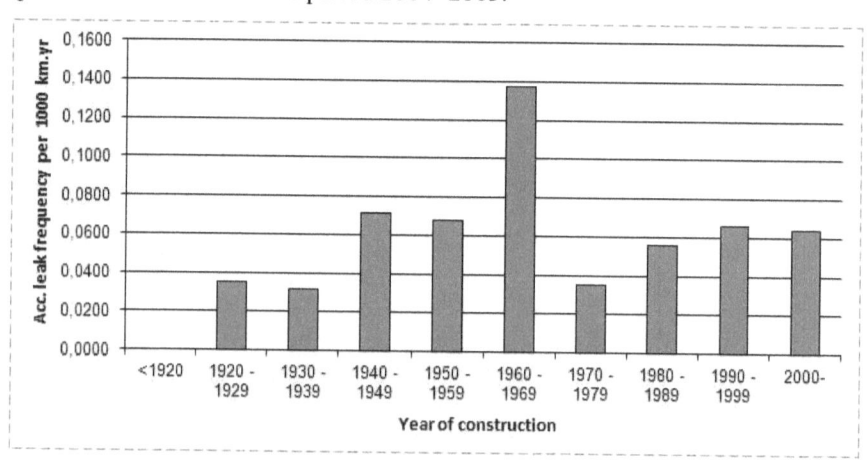

Figure A.3-2 – Relationship between accidental leak frequency of OP due to "Construction /material defect" and year of construction (2004–2013)

Considering the data in Figure A.3-2, it is difficult to recommend values of appropriate correction factor. Therefore, the accidental leak frequency of OP by the cause class "Construction /material defect" in calculation is taken equal to the average frequency (Table A.3-2 or A.3-3).

A.3.6 Accidental leak cause class: "Corrosion"

According to PHMSA data (Appendix 2, Section A.2.1, [12]) corrosion is the major cause of accidents related to oil leaks. As shown in Table A.3-1, accidental leak frequency for the period 1984–2013 is 0.27 cases per 1,000 km·yr. It is known that 59 % of all cases of accidents by the cause class "Corrosion" were caused by external corrosion, 40 % – by internal corrosion, the type of corrosion was unknown or not specified for 1 % of accidental leak frequency of OP. Therefore, it should be noted that the calculation, shown below, is primarily focused on accounting of implemented safety measures related to external corrosion (such as wall thickness and the presence / absence of protection means), since internal corrosion mainly depends on the quality of transported product (the presence of sulfur, salts, water, mechanical impurities, gases, etc.). None of the available statistical bases will allow making such calculation algorithm that would be able to take into account the content of certain impurities and their quantity in the transported product. However, the problem of internal corrosion of OP is not of much priority, since prepared oil is already transported through them unlike gathering pipelines, to which this problem is of high priority. Solution to this problem can be found when performing similar analysis of accidents precisely for gathering pipelines and/or requiring performance of special analysis in co-operation with experts in this field that will enable assigning of corresponding design factors.

Pipe wall thickness

Corrosion damages are mainly typical for thin-walled piping (wall thickness less than 10 mm) as shown in Table A.3-9. The data present distribution of number of accidents by the cause class "Corrosion" depending on the wall thickness, since it is impossible to determine the exposure of OP per wall thickness.

Table A.3-9 – Number of accidents due to the cause class "Corrosion" per pipe wall thickness (1984-2013)

Pipe wall thickness, mm	0-5	5-10	10-15	15-20
Number of accidents due to the cause class "Corrosion"	91	455	3	2

The wall thickness of pipelines certainly does not have any influence on the presence / absence of corrosion and on the rate of its development. However, it is obvious that the smaller is this value, the shorter is time of corrosion development that eventually will lead to accidental leak of OP, so it is easier to detect in a timely manner corrosion of OP with bigger wall thickness, when performing regular in-line inspection. According to Table A.3-9 it is obvious that in OP with wall thickness exceeding 10 mm the accidents by cause class "Corrosion" practically do not occur.

The proportion of accidents of OP with wall thickness over 10 mm of the total amount (573 accidents) (Appendix 2, Table A.2-4) is 0.009, whereas the portion of OP with wall thickness less than 10 mm – 0.953 (0.038 – portion of accidents with unknown wall thickness of OP).

Given the assumptions specified in the assignment of analogous accidental leak frequency reduction factor by the cause "External interference" (part A.3.4), the corresponding factor (k_{WT}) for OP with wall thickness over 10 mm may be taken equal to 0.018, and for OP with wall thickness less than 10 mm – 1[7].

Corrosion protection

OP operators take various protective measures to resolve the problem of corrosion. These measures e.g. include the use of cathodic protection and special corrosion-resistant coatings of piping (Figure A.3-3). Regular in-line inspections enable corrosion detection at early stage. PHMSA data [12] imply that the recently taken measures (Figure A.3-3) and improved prevention lead to long-term positive effect.

---

[7] Since 95 % of accidents, accounted for OP accidental leak primary frequency calculation by the cause class "Corrosion", took place at OP with wall thickness less than 10 mm, it seems appropriate do not perform frequency correction of such pipelines, therefore $k_{WT} = 1$.

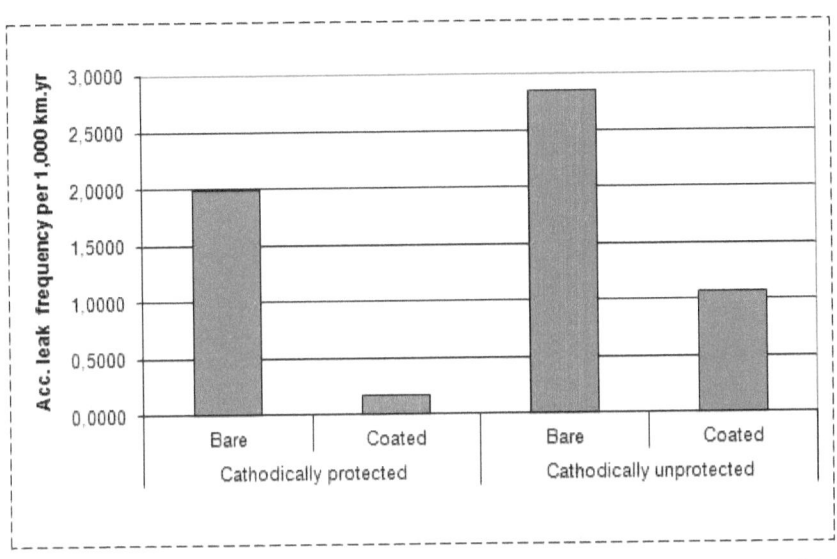

Figure A.3-3 – Relationship between corrosion protection measures and OP accidental leak frequency due to "Corrosion" (1984-2013)

Figure A.3-3 shows that cathodic protection combined with protective anti-corrosion coating are the most effective. This combination of protection measures provides reducing of accidental leak frequency of OP by the cause class "Corrosion" compared with the average frequency (Table A.3-1) that can be accounted with the factor 0.6. The accidental leak frequency of OP increases in comparison with the average frequency by 7.4 times, when using only cathodic protection without protective coating of pipeline. The accidental leak frequency of OP increases in comparison with the average frequency by 10.6 times without using any sort of protective measures. Availability of only corrosion protective coating of OP without the use of cathodic protection increases accidental leak frequency of OP in comparison with average frequency by 4 times.

Year of construction

Available data do not allow tracing of clear mathematical relationship of OP accidents by the cause class "Corrosion" to date of commissioning (Figure A.3-4) that makes impossible assignment of appropriate correction factor for the accidental leak frequency of OP calculation by the cause class "Corrosion".

Conservatively it can be assumed that accidental leak frequency of OP by the cause class "Corrosion" does not depend on the date of commissioning of OP at all, since it is assumed that the decrease in corrosion influence was not caused by the fact of "commissioning period" itself but by those protective measures (quality of pipes coating, in-line inspection, etc.) that are used in modern OP. In such a manner it is eliminated the effect of "double counting" of correlation factors "date of commissioning" /"protective measures" on an accident caused by corrosion [3].

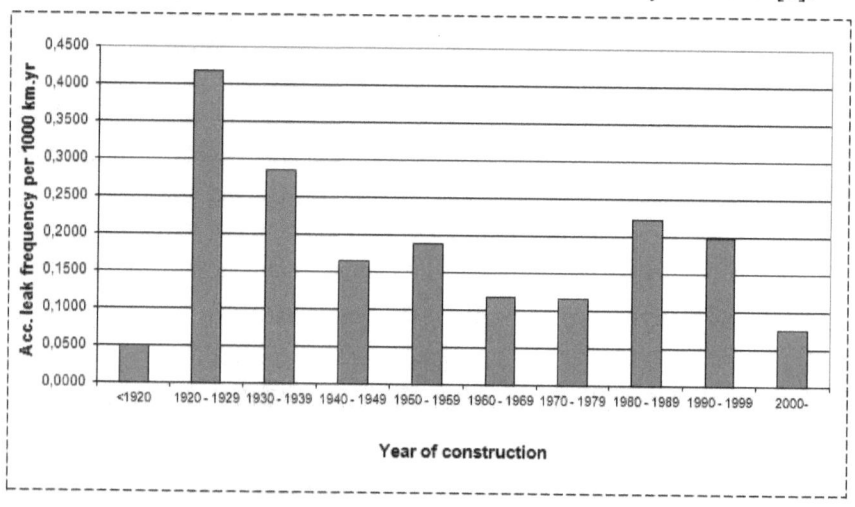

Figure A.3-4 – Relationship between OP year of construction and accidental leak frequency due to "Corrosion" (1984-2013)

Accidental leak frequency calculation of OP by the cause class "Corrosion"

Thereby, the accidental leak frequency of OP by the cause class "Corrosion" is calculated by the formula:

$$f_C = f_{PC} \cdot k_{WT} \cdot k_{CP}, \qquad (A.3\text{-}6)$$

where f_C – OP accidental leak frequency by the cause class "Corrosion";

f_{PC} – Primary frequency of OP accidental leak by the cause class "Corrosion" (Table A.3-2 or A.3-3);

k_{WT} – correction factor of OP accidental leak frequency by the cause class "Corrosion", taking into account the effect of OP wall thickness: k_{WT}=0,018 with OP wall thickness more than 10 mm; k_{WT} = 1, with OP wall thickness less than or equal to 10 mm;

k_{CP} – correction factor of OP accidental leak frequency by the cause class "Corrosion", taking into account the use of corrosion protection measures: k_{CP}=0,6 – if cathodically protected with protective coating; k_{CP}=7,4 – if cathodically protected without protective coating; k_{CP}=10,6 – if cathodically unprotected without protective coating; k_{CP}=4 – if cathodically unprotected with protective coating.

A.3.7 Accidental leak cause class: "Natural forces"

As follows from PHMSA data (Appendix 2, Part A.2.1, [12]), the accidental leak frequency of OP by the cause class "Natural forces" for the period 1984–2013 is 0.025 per 1,000 km·yr (Table A.3-1).

It is known that the direct causes of accidents were: Earthquakes – 22.2 %; Subsidence – 22.2 %; Washouts, floods – 16.7 %; Landslides – 1.9 %; Lightning – 1.9 %; Other and unknown – 35.2 %.

Particular attention should be paid to Natural forces in areas with unstable earth surface – i.e., in areas subjected to floods, landslides, earthquakes and hurricanes [3].

Decisive influence on the accidental leak frequency by the cause class "Natural forces" is caused by specifics of terrain with the pipeline route. Therefore, when determining the accidental leak frequency of OP caused by ground movement, the factors should be taken into account proposed for GP [3]: general seismicity; active tectonic faults; areas of soil liquefaction; landslide and mudslide areas; swamps and swampy areas; water crossings.

Analysis of the main geo hazards along OP route is similar to the analysis of GP [3], since the impact of natural forces on the route of cross-country pipeline does not much depend on the transported product.

Background accidents caused by natural forces

In order to avoid "double counting" in the calculation of expected accidents frequency per the same causes, that caused ground displacement, in "background" accidents should be left only those causes that cannot be explicitly taken into account in the analysis of specific geo hazards typical for the route. Therefore, as the "background" accidents frequency caused by natural forces, only accidents will be considered caused by "other and unknown causes" that compose, in conformity with PHMSA data, 35.2 % of the overall accidental leak frequency by the cause class "Natural forces" (Table A.3-2 or A.3-3).

The level of background accidents depends on OP parameters, in particular on pipeline diameter and pipe wall thickness.

Background accidents dependence on pipe diameter

Data on the level of accidents by the cause class "Natural forces" depending on pipeline diameter are shown in Figure A.3-5.

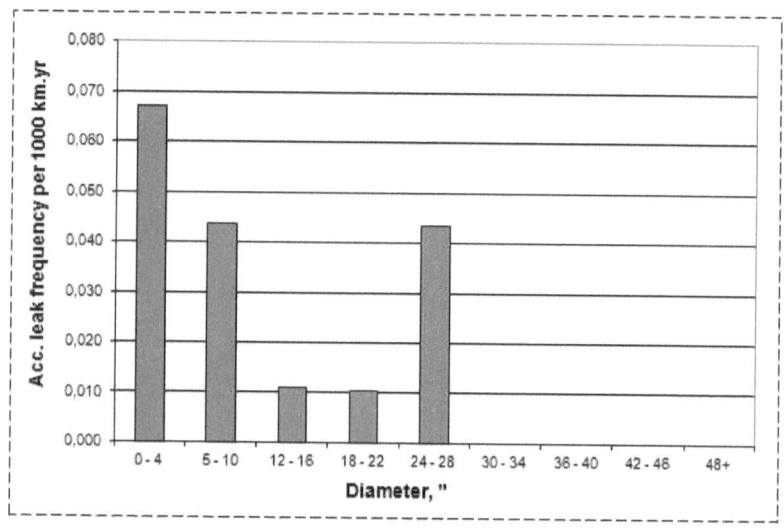

Figure A.3-5 – Relationship between OP diameter and accidental leak frequency due to "Natural forces" (1984-2013)

The diagram shown in Figure A.3-5 does not allow determining the precise mathematical dependence of the accidental leak frequency of OP by the cause class "Natural forces" on the pipeline diameter.

Background accidents dependence on pipeline wall thickness

It is obvious that the larger is wall thickness, the better is tube protection against any mechanical damage, including resulting from natural forces (ground movement), confirmed by Table A.3-10.

Table A.3-10 – Number of accidents due to the cause class "Natural forces" per pipe wall thickness (1984-2013)

Pipe wall thickness, mm	0-5	5-10	10-15	>15
Number of accidents due to the cause class "Natural forces"	0	44	6[8]	0

Proportion of accidents at OP with wall thickness 10 mm of the total amount (54 accidents) (Appendix 2, Table A.2-4) is 0.11, whereas OP with wall thickness less than 10 mm – 0.81 (0.07 – percentage of accidents with unknown wall thickness of OP).

With account of assumptions specified in the appointment of similar accidental leak frequency reduction factor of OP by the cause class "Natural forces" (Part A.3.4), the corresponding background accidental leak frequency reduction factor of OP (k_{WT}) with wall thickness over 10 mm can be taken 0.22, for OP with wall thickness less than 10 mm – 1[9].

Influence of overall seismicity on the accidental leak frequency of OP by the cause class "Natural forces"

If the OP route belongs to seismically active zone, then the influence of overall seismicity on the frequency of accidents caused by natural forces should be considered in calculations as well. All pipeline constructions are designed with

---

[8] The data appeared related to the pipe wall thickness for some previously recorded accidents by the cause class "Natural forces" in the summary data on accidents for the period 2010–2014, therefore these tables differ significantly from the previously published data [1,2].

[9] Since 80 % of accidents, accounted for OP accidental leak primary frequency calculation by the cause class "Natural forces", took place at OP with wall thickness less than 10 mm, it seems appropriate do not perform frequency correction of such pipelines, therefore $k_{WT} = 1$.

account of probable earthquakes with peak ground acceleration for strength-level event (SLE) and ductility-level event (DLE) [15]. Thereby, the pipelines are calculated in such a way as to eliminate the possibility of leak at any earthquakes of DLE level. The peak ground acceleration (PGA), corresponding to DLE earthquake for each region / area of pipeline routing, can be determined by conducting separate study for seismic hazard in this region/area assessment. Experts suggested [3] that an earthquake with PGA exceeding twice the PGA value typical for this region/area of pipeline route, can destroy the pipeline. For example, in the framework of the project "Sakhalin-II" the whole range of scientific research was held to assess the seismic hazard of the Trans-Sakhalin pipeline system routing area. As a result of this study it was found that PGA level in the southern part of Sakhalin Island does not exceed 0.73g. Therefore, it was noted [3] that an earthquake with PGA equal to 1.4g may destroy the OP. Also based on the procedure data of probabilistic seismic hazard analysis (PSHA) it was found that an earthquake with a return period of 10,000 years or more corresponds to PGA level 0.73g at different points along the pipeline route. It is also assumed that the pipeline will be destroyed by such an earthquake in one place per 10 km route (i.e. in the section equal to several lengths of seismic waves). Then the assessment of pipeline destruction frequency due to earthquake load beyond design basis (extremely high) seismic loads is performed by calculating the PGA probability of exceeding the level 1.4g in a given point subject to one destruction per 10 km of the route by this earthquake [3]:

$$f_{SA}=1/\{T(PGA=1.4g)\cdot 10\}, \qquad (A.3-7)$$

where f_{SA} – accidental leak frequency of OP (for normal safety class) caused by beyond design seismic action,

$T(PGA=1.4g)$ – frequency of earthquakes recurrence, characterized by excess of the 1.4 g level of PGA.

The formula A.3-7 may be presented in the general form:

$$f_{SA}=1/\{T(2PGA)\cdot 10\}, \qquad (A.3-8)$$

where $T(2PGA)$ – frequency of earthquakes recurrence, characterized by double excess of the level of peak ground acceleration (PGA) typical for analyzed region / area of a pipeline route.

Dependence of accidental leak frequency of OP due to seismic ground movement on the class of pipeline safety

It is noted [3] the need to account in calculations the dependence of accidental leak frequency caused by seismic ground movement, on the class of pipeline safety (wall thickness). Therewith, according to PHMSA data [12] it is impossible to identify clear mathematical dependence of accident frequency caused by seismic action on the class of OP safety. It can be assumed that dependence of the frequency of OP damage on the wall thickness in case of seismic action has the same functional relationship, as in the case of external interference [3]. Possible substantiation of this fact – in both cases we are dealing with short-term external power load on a pipeline. Therefore, for accountability of the safety class of pipeline it is proposed to multiply the value of frequency of pipeline destruction resulted by seismic loads (beyond design basis, see above) by correction factor (k_{WT}), which for pipelines with wall thickness over 10 mm can be taken equal to 0.022. For OP with wall thickness less than 10 mm the frequency reduction factor is taken equal to 1 (Section A.3.4).

Active tectonic faults crossings

The project of modern OP for each active tectonic fault crossing shall provide special technical measures that take into account specific characteristics of a specific fault; therefore, we may assume that any ground movements at the fault, caused by the earthquake of DLE level, do not lead to the pipeline destruction.

Nevertheless, on the basis of special investigations in order to assess the seismic hazard in the region/area of pipeline route it is possible to set the earthquake recurrence period, causing OP damage in the area of tectonic fault. E.g., the expert assessments were taken into account in [3] related to earthquake recurrence period, causing GP damage in segment of tectonic fault in conditions of the Southern Sakhalin. It was shown that using conservative methods, in the worst case the recurrence of such an earthquake will be not less than 30,000 years. Therefore, the frequency of GP destruction at the active tectonic fault was taken equal to the primary frequency value – $3.3 \cdot 10^{-5}$ 1/year [3, 5].

The estimated length of active tectonic fault crossing segment shall be equal to the size of fault uncertainty zone (in [3, 5] it was 180 m).

Accidental leak frequency of OP caused by landslides and mudslides

In sections of the OP route, where landslides and mudslides are possible, the design engineering solutions should be provided, which ensure its tightness under normal operating conditions.

Taking into account the data [3], we take also for OP the accidental leak frequency caused by landslides and mudslides (f_L) equal to 10 % of the accidental leak frequency caused by seismic action (f_{SA}) in this section of the route. Since it cannot be excluded that strong earthquake may cause landslides and mudslides in these areas with higher intensity of impact on OP, than it was designed in the project, it may cause the pipeline damage. Therefore, in this case it is assumed the mechanism of seismically activated ground movement, causing hazardous secondary processes (landslides, mudslides), which lead to OP accidents.

Accidental leak frequency of OP caused by possible seismic soil liquefaction

Special engineering solutions should be developed in OP projects for areas with possible seismic soil liquefaction to ensure its tightness under normal operating conditions.

Strong earthquake may cause soil liquefaction in these areas with loads exceeding the estimated value that may cause the pipeline damage. The accidental leak frequency of OP caused by seismic soil liquefaction (f_{LF}) should be defined for such events. Therefore, for OP the accidental leak frequency caused by seismic soil liquefaction (f_{LF}) can be taken equal to 10 % of the accidental leak frequency caused by seismic action (f_{SA}) in this section of the route [3].

Accidental leak frequency of OP caused by soil movement at OP water crossings

It does not seem possible to estimate the frequency of accidents occurred at the river crossings on the basis of PHMSA data [13]. Moreover, reports of CONCAWE [6] group do not contain relevant information about changing the intensity of accidents at the OP route crossing water obstacles. Therefore, to estimate the frequency of accidents at the water crossings the data given for GP in [3] may be used. According to these data the accidental leak frequency of OP at water crossings (f_{WC}), caused by the ground movement (due to washout), is taken 5 times higher than the background frequency (f_{BNF}) (except crossings with HDD method performance). The length of water crossing section can be conservatively taken equal to the width of profile of the watercourse washout limit for the 30-year period.

Accidental leak frequency of OP caused by ground movement at swamp and swampy areas crossings

The accidental leak frequency of OP caused by ground movement at swamp and swampy crossings (f_S) in conformity with [3] is taken 2 times higher than the background frequency (f_{BNF}). The length of the swampy crossing section is taken equal to the width of the swampy crossing.

It should be noted that at the crossings with directional drilling performance, in consequence of the great depth of cover of OP, when calculating the accidental leak frequency of OP caused by ground movement due to natural force, the effect of water, swamps, soil liquefaction, landslides and mudslides crossings is not taken into account.

Accidental leak frequency calculation of OP by the cause class "Natural forces"

The accidental leak frequency of OP caused by natural force (ground movement), is calculated by the formula:

$$f_{NF} = f_{BNF} + f_{SA} + f_{TF} + f_L + f_{LF} + f_{WC} + f_S, \qquad (A.3-9)$$

where f_{BNF} - background accidental leak frequency of OP by the cause class "Natural forces" caused by geological hazards of an OP route that are not accounted in the basic formula, defined by taking into account safety classes of pipe: k_{WT}=0,22, if the OP wall thickness is more than 10 mm; k_{WT}=1, if the OP wall thickness is less than or equal to 10 mm;

f_{SA} - accidental leak frequency of OP caused by seismic action (formula A.3-8), defined by taking into account safety classes of pipe: k_{WT}=0,022, if the OP wall thickness is more than 10 mm; k_{WT}=1, if the OP wall thickness is less than or equal to 10 mm;

f_{TF} - accidental leak frequency of OP caused by its destruction on the active tectonic fault;

f_L - accidental leak frequency of OP caused by landslides and mudslides;

f_{LF} - accidental leak frequency of OP caused by seismic soil liquefaction;

f_{WC} - accidental leak frequency of OP caused by soil movement at OP water crossings;

f_S - accidental leak frequency of OP at swampy areas crossing.

The number of components in the formula A.3-9 varies, depending on the presence of one or another geo hazard in the analyzed section of OP.

A.3.8 Accidental leak cause class: "Incorrect operation"

As follows from Table A.3-1, the accidental leak frequency of OP by the cause class "Incorrect operation" for the period 1984–2013 is 0.011 cases per 1,000 km·yr.

According to PHMSA data [12], accidents caused by "human factor" are mainly observed at the OP with diameter up to 34", as shown in Figure A.3-6.

Noted clear dependence of accidental leak frequency of OP on pipeline diameter can be described by the following regression equation:

$$f_{IO} = -0.019 \cdot \ln(D) + 0.0681 \qquad (A.3\text{-}10)$$

Therefore, Table A.3-11 shows results of frequency reduction factor calculation for various diameters of OP with account of dependence A.3-10.

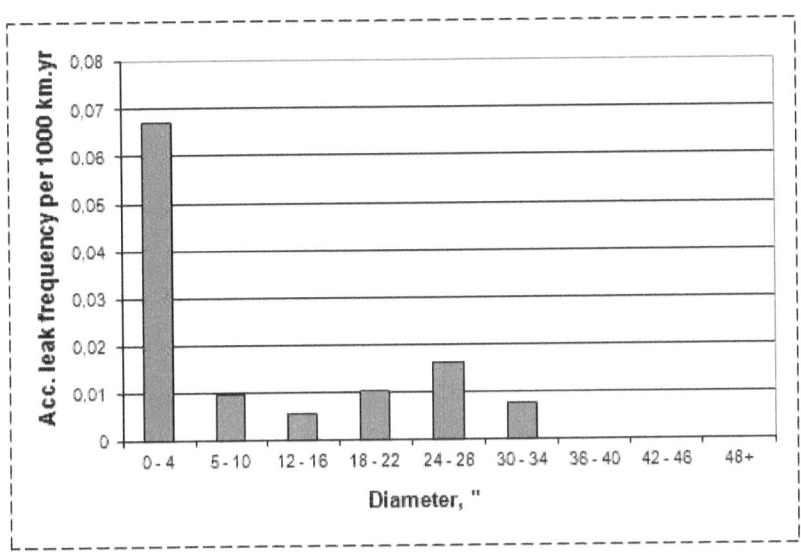

Figure A.3-6 – Relationship between OP diameter and accidental leak frequency due to "Incorrect operation" (1984-2013)

Table A.3-11 – Accidental leak frequency reduction factors of OP by the cause class "Incorrect operation" depending on pipeline diameter

Diameter, "	Primary accidental leak frequency, 1000 km.yr	Accidental leak frequency per dependence A.3-10, 1000 km.yr	k_D
30		0,0035	0,308
20		0,0112	0,990
14	0,0113	0,0180	1,590
8		0,0286	2,531
4		0,0418	3,696

Accidental leak frequency calculation of OP by the cause class "Incorrect operation"

The accidental leak frequency of OP by the cause class "Incorrect operation" is calculated by the formula:

$$f_{IO} = f_{PIO} \cdot k_D, \qquad (A.3-11)$$

где

f_{IO} – OP accidental leak frequency by the cause class "Incorrect operation";

f_{PIO} – Primary frequency of OP accidental leak by the cause class "Incorrect operation" (Table A.3-2 or Table A.3-3);

k_D - Correction factor of OP accidental leak frequency by the cause class "Incorrect Operation", taking into account the effect of OP diameter (taking into account the dependence A.3-10 or in Table A.3-11).

A.3.9 Accidental leak cause class: "Other and Unknown"

110 accidents occurred as a result of cause different from accepted classification were registered during 1984–2013 in PHMSA database (Appendix 2, Section A.2.1, [12]), forming the accidental leak frequency of OP equal to 0.052 cases per 1,000 km·yr (Table A.3-1).

The accidental leak frequency of OP by the cause class "Other and unknown" is taken in calculations as constant and equal to the primary frequency according to Table A.3-2 and Table A.3-3.

APPENDIX 4.
AN EXAMPLE OF ACCIDENTAL LEAK FREQUENCY OF OP CALCULATION

As an example we present results of calculation, using the proposed procedure for some sections of new OP commissioned as part of one of the oil and gas projects in Sakhalin Island.

The OP route location has high complexity due to the following: high seismic area; presence of active tectonic faults, large number of water courses, swamps and swampy areas, geo hazards, presence of engineering structures and utilities. To ensure an adequate safety level in conformity with the best world practice, it was provided the wide range of measures in the project to reduce the risk of OP accidents.

In particularly vulnerable sections of the OP route, the project provided and implemented the use of pipes of different safety classes, characterized by pipe wall thickness. The wall thicknesses of the considered section of OP with diameter 24" (610 mm) depending on the safety class are as follows: 9.5 mm – normal; 11.4 mm – medium; 13.7 mm – high; 19.1 mm – seismic.

Electrochemical protection of OP provides permanent cathodic protection using several cathodic protection stations installed along the route. Covers protection of roads and railways crossing is performed by extended magnesium anodes.

OP has a three-layer polyethylene anticorrosion factory coating, which minimum thickness is 3.2 mm.

When constructing OP water crossings, trench and trenchless methods (method of horizontal directional drilling) were used, special measures were implemented to ensure the sustainability of the designed pipeline position against the floating up in streambeds, floodplains and swamps.

Minimum OP depth cover is taken equal to 1 m to the upper generatrix. At railways and roads crossings, it is laid in protective cases (housings) made of steel pipes. Special project solutions were designed for active tectonic fault crossings that provide leak-tightness of OP at ductility-level event (e.g. thick-walled pipes, the optimum angle of the fault crossing, etc.). The wide range of special design solutions was provided for crossings areas of slope processes evolution, as well as seismological and geodetic monitoring of OP routs in operation.

Table A.4-1[10] shows results of the expected accidental leak frequency calculation of OP 24" for one of the most difficult sections with active tectonic faults (3 pcs.), soil liquefaction areas, landslides, swamp crossings, water courses and roads. Each table row represents section of the OP route, made of pipeline of certain safety class. The beginning of each section (kilometer point of route) is shown in the leftmost column of the table. The next columns show the length (in meters) of a given interval of directional drilling section, soil liquefaction, landslides, shears, swamps, water courses and roads. Safety classes are also listed of each pipeline section. Calculated accidental leak frequencies of OP for Rupture cases and leaks are shown in the right side of Table A.4-1 in two outermost columns. Frequencies calculated for the period 2004–2013 with account of small leaks (Table 1, (3); Appendix 3, Table A.3-3) were taken as primary frequencies in calculations.

Results of calculations are presented in graphical form in Figure A.4-1. It is seen that the expected frequency of leaks about one order exceeds the frequency of OP full ruptures. The amplitude of accidental leak frequency variation along the OP route amounts up to 3 times. However, the expected accidental leak frequency of OP with full rupture (average value – 0.063 1/1,000 km·yr) is slightly lower than the average in Russia for the period 2008–2012 (average intensity of accidents – 0.075–0.08 1/1,000 km·yr [16]).

Furthermore, calculated values of accidental leak frequencies without including changes in pipeline wall thickness depending on OP section are presented in Figure A.4-1 for comparison. It is clearly shown significant impact caused by special safety measures implementation on expected accidental leak frequencies of OP.

Generally, when designing cross-country pipelines, additional activities and their impact on reliability (resilience) of pipeline shall be chosen "with reserve" providing

---

[10] For convenience of presentation Table A.4-1 presented in abbreviated form, some sections were excluded of considered OP (lines marked with "...") for which the calculation was performed.

ultimately the resulting accidental leak frequencies in hazardous areas often even smaller than in "ordinary" pipeline sections.

Figure A.4-2 also shows the values of accidental leak frequency of OP by cause class "Natural forces" with account of pipe wall thickness and without. These two diagrams show the impact of safety measures (pipeline wall thickness) on calculated values of expected accidental leak frequency of OP. However, despite taken safety measures at locations of active tectonic faults crossings by OP route (three clearly defined peaks in Figure A.4-2, where frequencies of two diagrams coincide) the expected value of accidental leak frequency of OP is significantly higher than in adjacent sections. The average frequency in the right side of diagrams is significantly higher than in other parts of considered section of OP. This is due to the fact that the OP route crosses swamps in this area.

In general, Figure A.4-1 and A.4-2 show "sensitivity" of the proposed procedure to safety measures provided by the project and to variety of conditions of OP routing.

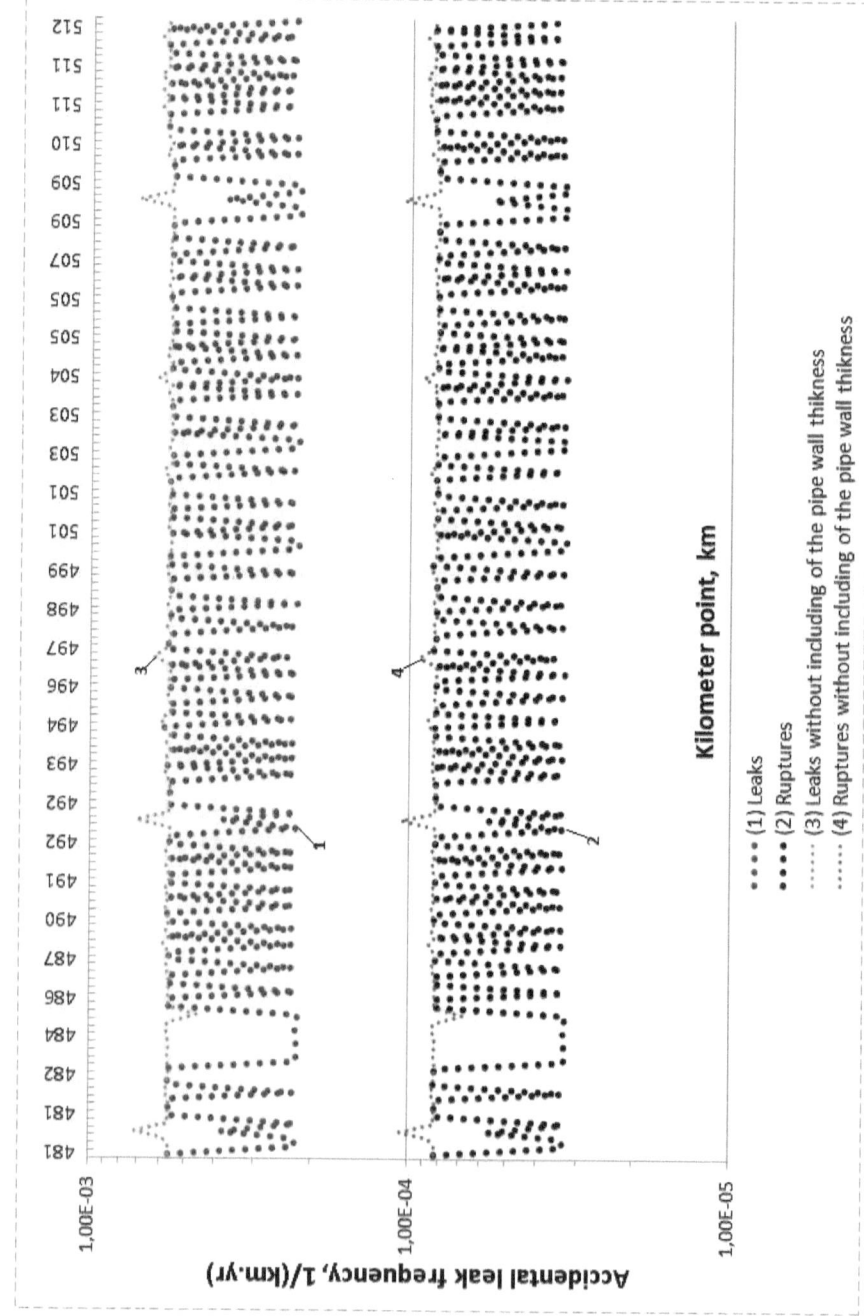

Figure A.4-1 – Calculation results of expected accidental leak frequency of OP 24" (610 mm)

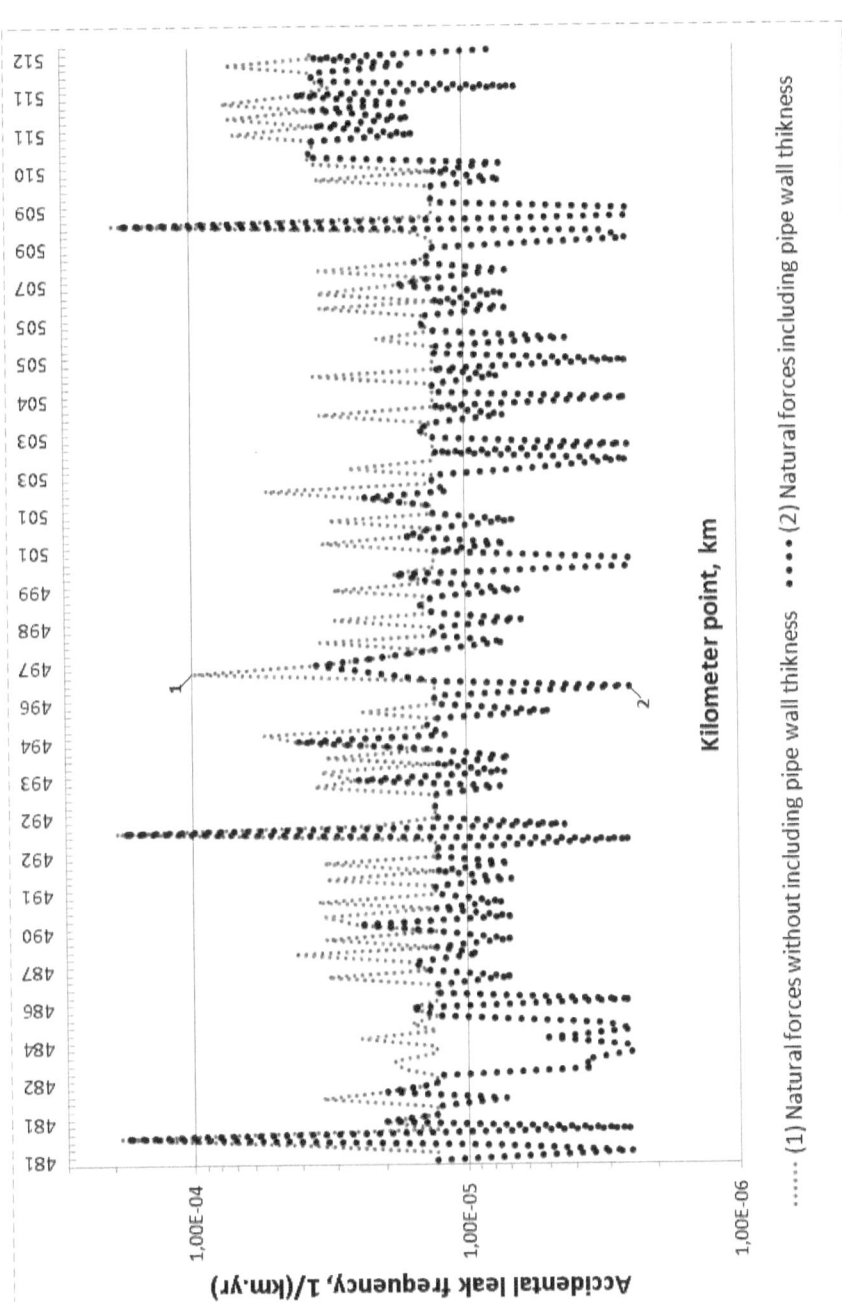

Figure A.4-2 – Calculation results of expected accidental leak frequency of OP 24" (610 mm) due to "Natural forces"

Table A.4-1 – Calculation results of expected accidental leak frequency of OP 24" (610 mm)

Section start, km	HDD crossing, m	Area features, m				Safety class, m				Water crossings, m		Roads crossings, m	Acc. leak frequency, 1/km.yr	
		Fault, available/n|a	Liquefaction	Landslide	Mudslides	Normal	Medium	High	Seismic	Swamps	Rivers		Leaks	Ruptures
480,745	-	-	-	-	-	192,31	-	-	-	-	-	-	5,59E-04	8,25E-05
480,936	-	-	-	-	-	-	11,30	-	-	-	-	-	2,22E-04	3,27E-05
480,948	-	-	-	-	-	-	-	53	-	-	22,4	-	2,26E-04	3,34E-05
481	-	1	-	-	-	-	-	185	-	-	-	-	3,82E-04	5,63E-05
481,185	-	-	-	-	-	-	11,30	-	-	-	-	-	2,22E-04	3,27E-05
481,196	-	-	-	-	-	157,36	-	-	-	-	20,3	-	5,65E-04	8,34E-05
481,345	-	-	-	-	-	555	-	-	-	-	-	-	5,59E-04	8,25E-05
481,9	-	-	-	-	-	89,27	-	-	-	-	-	-	5,59E-04	8,25E-05
481,989	-	-	-	-	-	-	56,5	-	-	-	22	-	2,26E-04	3,33E-05
482,045	-	-	-	-	-	161,96	-	-	-	-	20,3	-	5,65E-04	8,34E-05
...	...	...	...	...	...	...	...	...	...	...	...	...	...	...
482,836	-	-	-	-	-	-	988,7	-	-	150	20,3	-	2,23E-04	3,29E-05
483,822	-	-	-	-	-	-	-	277,85	-	-	28,8	-	2,23E-04	3,29E-05
484,1	-	-	-	-	-	-	-	56,26	-	-	-	-	2,22E-04	3,27E-05
...	...	...	...	...	...	...	...	...	...	...	...	...	...	...
485,1	150	-	-	-	-	-	877	-	-	120	-	-	2,20E-04	3,25E-05
485,95	-	-	-	-	-	14	-	-	-	-	-	-	5,59E-04	8,25E-05
485,964	-	-	-	-	-	337,23	-	-	-	-	20,3	-	5,62E-04	8,29E-05
...	...	...	...	...	...	...	...	...	...	...	...	...	...	...
487,21	-	-	-	-	-	-	56,5	-	-	-	20,2	-	2,26E-04	3,33E-05
487,267	-	-	-	-	-	1248,4	-	-	-	-	20,3	-	5,60E-04	8,26E-05
488,5	-	-	-	-	-	376,18	-	-	-	-	20,3	-	5,62E-04	8,29E-05
488,875	-	-	-	-	-	-	113	-	-	-	61,3	-	2,28E-04	3,36E-05
488,988	-	-	-	-	-	665,26	-	-	-	-	-	-	5,59E-04	8,25E-05
489,64	-	-	-	-	-	-	56,5	-	-	-	20,3	-	2,26E-04	3,33E-05
489,697	-	-	-	-	-	311,12	-	-	-	-	-	-	5,59E-04	8,25E-05
490	-	-	-	-	-	91,3	-	-	-	-	20,3	-	5,70E-04	8,41E-05
490,088	-	-	-	-	-	-	56,5	-	-	-	20,5	-	2,26E-04	3,33E-05
490,144	-	-	-	-	-	759,94	-	-	-	-	-	-	5,59E-04	8,25E-05
490,896	-	-	-	-	-	-	56,5	-	-	-	22,6	-	2,26E-04	3,34E-05

Section start, km	HDD crossing, m	Area features, m					Safety class, m			Water crossings, m		Roads crossings, m	Acc. leak frequency, 1/km.yr	
		Fault,	Liquefaction	Landslide	Mudslides	Normal	Medium	High	Seismic	Swamps	Rivers		Leaks	Ruptures
490,953	-	-	-	-	-	557,12	-	-	-	-	-	-	5,59E-04	8,25E-05
491,5	-	-	40	-	-	224,19	-	-	-	-	-	-	5,59E-04	8,25E-05
491,717	-	-	-	-	-	-	56,63	-	-	-	20,4	-	2,26E-04	3,33E-05
491,774	-	-	-	-	-	45,39	-	-	-	-	-	-	5,59E-04	8,25E-05
491,817	-	-	-	-	-	-	56,17	-	-	-	20,5	-	2,26E-04	3,33E-05
...	...	...	...	...	...	...	...	...	...	...	...	...	...	...
492,134	-	1	-	-	-	-	-	126,05	-	-	-	-	3,82E-04	5,63E-05
492,259	-	-	-	-	-	-	158,2	-	-	-	23,6	-	2,23E-04	3,30E-05
...	...	...	...	...	...	...	...	...	...	...	...	...	...	...
493,146	-	-	-	-	-	-	56,5	-	-	-	23,5	-	2,26E-04	3,34E-05
493,202	-	-	-	-	-	798,42	-	-	-	390	20,3	-	5,70E-04	8,41E-05
493,996	-	-	-	-	-	-	56,5	-	-	-	21,5	-	2,26E-04	3,33E-05
494,053	-	-	-	-	-	133,98	-	-	-	-	-	-	5,59E-04	8,25E-05
494,188	-	-	-	-	-	-	56,5	-	-	-	20,5	-	2,26E-04	3,33E-05
494,242	-	-	-	-	-	57,88	-	-	-	-	-	-	5,59E-04	8,25E-05
494,3	-	-	-	-	-	166,5	-	-	-	166	20,3	-	5,84E-04	8,62E-05
494,466	-	-	-	-	-	-	56,5	-	-	56	20,8	-	2,30E-04	3,39E-05
494,522	-	-	-	-	-	1005,2	-	-	-	-	20,3	-	5,60E-04	8,27E-05
495,526	-	-	-	-	-	61,07	-	-	-	-	-	-	5,59E-04	8,25E-05
495,587	-	-	-	-	-	-	135,6	-	-	-	26,35	-	2,24E-04	3,30E-05
495,722	-	-	-	-	-	77,88	-	-	-	-	-	-	5,59E-04	8,25E-05
495,8	-	-	-	-	-	528,22	-	-	-	-	-	20	5,63E-04	8,31E-05
496,325	-	-	-	-	-	-	33,9	-	-	-	-	-	2,22E-04	3,27E-05
496,359	-	-	-	-	-	571,7	-	-	-	-	-	-	5,59E-04	8,25E-05
496,929	-	-	-	-	-	-	56,5	-	-	166	21,8	-	2,38E-04	3,52E-05
496,986	-	-	-	-	-	314,28	-	-	-	314	-	-	5,78E-04	8,53E-05
...	...	...	...	...	...	...	...	...	...	...	...	...	...	...
501,415	-	-	200	-	-	385,11	-	-	-	-	-	-	5,59E-04	8,25E-05
501,8	-	-	-	-	-	107,43	-	-	-	-	20,3	-	5,68E-04	8,39E-05
501,906	-	-	-	-	-	-	56,5	-	-	-	41,66	-	2,30E-04	3,39E-05
501,962	-	-	-	400	-	690,27	-	-	-	-	-	-	5,59E-04	8,25E-05
...	...	...	...	...	...	...	...	...	...	...	...	...	...	...
504,138	-	-	-	-	-	-	56,5	-	-	-	21,6	-	2,26E-04	3,33E-05

73

Section start, km	HDD crossing, m	Area features, m				Safety class, m				Water crossings, m		Roads crossings, m	Acc. leak frequency, 1/km.yr	
		Fault	Liquefaction	Landslide	Mudslides	Normal	Medium	High	Seismic	Swamps	Rivers		Leaks	Ruptures
504,195	-	-	-	-	-	204,43	-	-	-	-	-	-	5,59E-04	8,25E-05
504,391	-	-	-	-	-	-	33,9	-	-	-	-	20	2,23E-04	3,29E-05
...	...	...	...	...	...	...	...	...	...	...	...	...	...	...
505,328	-	-	-	-	-	72,32	-	-	-	-	-	-	5,59E-04	8,25E-05
505,4	-	-	50	-	-	806,53	-	-	-	-	20,3	20	5,63E-04	8,31E-05
506,2	-	-	-	-	-	1037,2	-	-	-	-	20,3	20	5,62E-04	8,29E-05
507,222	-	-	-	-	-	117,2	56,5	-	-	-	21,1	-	2,26E-04	3,33E-05
507,278	-	-	-	-	-	-	56,5	-	-	-	-	-	5,59E-04	8,25E-05
507,391	-	-	-	-	-	-	56,5	-	-	-	21,5	-	2,26E-04	3,33E-05
507,448	-	-	-	-	-	255,48	-	-	-	-	22,4	-	5,63E-04	8,31E-05
507,7	-	-	-	-	-	218,64	-	-	-	-	-	-	5,59E-04	8,25E-05
507,91422	-	-	-	-	-	-	56,5	-	-	-	21,2	-	2,26E-04	3,33E-05
...	...	...	...	...	...	...	...	...	...	...	...	...	...	...
508,657	-	-	-	-	-	-	11,30	-	-	-	-	-	2,22E-04	3,27E-05
508,668	-	1	-	-	-	-	-	424,75	-	-	21,45	-	2,22E-04	3,28E-05
509,088	-	-	-	-	-	-	-	-	206,89	-	-	-	3,82E-04	5,63E-05
509,295	-	-	-	-	-	-	-	126,17	-	-	-	-	2,22E-04	3,27E-05
509,421	-	-	-	-	-	-	176,9	-	-	-	-	-	2,22E-04	3,27E-05
...	...	...	...	...	...	...	...	...	...	...	...	...	...	...
509,896	-	-	-	-	-	-	135,6	-	-	-	52,2	20	2,26E-04	3,34E-05
...	...	...	...	...	...	...	...	...	...	...	...	...	...	...
510,818	-	-	-	-	-	-	33,9	-	-	34	21	-	2,33E-04	3,43E-05
510,852	-	-	-	-	-	36,38	-	-	-	36	-	-	5,78E-04	8,53E-05
510,888	-	-	-	-	-	-	33,9	-	-	34	22,7	-	2,33E-04	3,44E-05
510,922	-	-	120	-	-	208,88	-	-	-	209	-	-	5,78E-04	8,54E-05
511,131	-	-	-	-	-	-	33,9	-	-	34	24,9	-	2,34E-04	3,45E-05
511,165	-	-	-	-	-	206,99	-	-	-	209	22	-	5,84E-04	8,61E-05
511,374	-	-	-	-	-	-	226	-	-	-	70,9	20	2,25E-04	3,33E-05
...	...	...	...	...	...	...	...	...	...	...	...	...	...	...
512,116	-	-	-	-	-	-	33,9	-	-	34	23,8	-	2,33E-04	3,45E-05
512,15	-	-	-	-	-	122,17	-	-	-	122	-	-	5,78E-04	8,53E-05
512,272	-	-	250	-	50	-	858,8	-	-	859	20,8	-	2,26E-04	3,34E-05

APPENDIX 5 (FOR REFERENCE).
BRIEF DESCRIPTION OF ANALYZED ACCIDENTS CONSEQUENCES: FIRES AND EXPLOSIONS

Analysis of consequences is not related to the accidental leak frequency calculation of OP. However, it seems appropriate to provide brief overview of accidents that caused fire or explosion (Table A.5-1, A.5-2).

Table A.5-1 – Accidents with the fire/explosion in the period of 1984-2013

Year	Consequences	Accident cause class	Location	Damage type
1984	Fire	External interference	Line pipe (body of pipe)	N/a
1985	Fire	External interference	Line pipe (body of pipe)	N/a
1985	Fire	External interference	Line pipe (body of pipe)	N/a
1985	Fire	Corrosion	Line pipe (body of pipe)	N/a
1985	Fire	Corrosion	Line pipe (body of pipe)	N/a
1986	Fire	Other and Unknown	Line pipe (body of pipe)	N/a
1986	Fire	External interference	Line pipe (body of pipe)	N/a
1986	Fire	External interference	Line pipe (body of pipe)	N/a
1987	Fire	Construction/material defect (weld defect)	Line pipe (welding fitting)	N/a
1987	Fire	Other and Unknown	Line pipe (body of pipe)	N/a
1987	Fire	External interference	Line pipe (scraper trap)	N/a
1987	Fire	Construction/material defect (pipe defect)	Line pipe (body of pipe)	N/a
1988	Fire	Construction/material defect (pipe defect)	Line pipe (body of pipe)	N/a
1989	Fire	Other and Unknown	Line pipe (girth)	N/a
1992	Fire	External interference	Line pipe (body of pipe)	N/a
1992	Fire	Corrosion	Line pipe (body of pipe)	N/a
1993	Fire	Incorrect operation	Line pipe (body of pipe)	N/a
1994	Fire	Natural forces (washout)	Line pipe (body of pipe)	N/a
1995	Fire /explosion	Natural forces (earthquake)	Line pipe (body of pipe)	N/a
1996	Fire	Natural forces	Line pipe (body of pipe)	N/a
1996	Fire	Other and Unknown	Line pipe (valve)	N/a
1997	Fire	External interference	Line pipe (body of pipe)	N/a
1999	Fire /explosion	Other and Unknown	Line pipe (valve)	N/a
2000	Fire	External interference	Line pipe (body of pipe)	N/a
2000	Fire	Other and Unknown	Other	N/a
2002	Fire	Construction/material defect (material failure and/or pipe seam)	Line pipe (pipe seam)	Tear/crack (longitudinal rupture)
2004	Fire /explosion	Construction/material defect (material failure and/or pipe	Line pipe (component)	Connection failure

Year	Consequences	Accident cause class	Location	Damage type
		seam)		
2005	Fire /explosion	External interference (excavation damage)	Line pipe (body of pipe)	Puncture
2007	Fire	External interference (nearby fire or explosion)	Line pipe (body of pipe)	Other
2007	Fire	Incorrect operation	Other	Other (flash fire during a cut out)
2007	Fire	Incorrect operation	Line pipe (bolted fitting)	Other
2010	Fire	Incorrect operation (overpressure)	Line pipe (body of pipe)	Tear/crack (longitudinal rupture)
2011	Fire	Incorrect operation	Line pipe (railroad crossing)	Inadvertent burn-through
2013	Fire	Incorrect operation	Line pipe (body of pipe)	Burn through of carrier pipe

Table A.5-2 – Number of accidents with the fire/explosion and cause classes in the period 1984–2013

Cause class	Number of Fire	Number of Explosion
Construction /material defect	5	1
Incorrect operation	6	-
Corrosion	3	-
Natural forces	3	1
External interference	11	1
Other and Unknown	6	1
Total	**34**	**4**

I want morebooks!

Buy your books fast and straightforward online - at one of the world's fastest growing online book stores! Environmentally sound due to Print-on-Demand technologies.

Buy your books online at
www.get-morebooks.com

Kaufen Sie Ihre Bücher schnell und unkompliziert online – auf einer der am schnellsten wachsenden Buchhandelsplattformen weltweit!
Dank Print-On-Demand umwelt- und ressourcenschonend produziert.

Bücher schneller online kaufen
www.morebooks.de

OmniScriptum Marketing DEU GmbH
Heinrich-Böcking-Str. 6-8
D - 66121 Saarbrücken
Telefax: +49 681 93 81 567-9

info@omniscriptum.com
www.omniscriptum.com

www.ingramcontent.com/pod-product-compliance
Lightning Source LLC
Chambersburg PA
CBHW031535210526
45464CB00003B/1020